Zoltan J Kiss

The Quantum Impulse and the Space-Time Matrix

*The power of the Hydrogen process
and the Pyramid*

2015

Order this book online at www.trafford.com
or email orders@trafford.com

Most Trafford titles are also available at major online book retailers.

Print information available on the last page.

ISBN: 978-1-4907-6192-3 (sc)
ISBN: 978-1-4907-6193-0 (e)

Library of Congress Control Number: 2015910579

Trafford rev. 07/08/2015

Trafford PUBLISHING® www.trafford.com

North America & international
toll-free: 1 888 232 4444 (USA & Canada)
fax: 812 355 4082

"I will prise thee with uprightness of heart,
when I shall have learned thy righteous judgements."
Psalm 119/7 Old Testament
"Open thou mine eyes, that I may behold wondrous
things out of thy law."
Psalm 119/18 Old Testament

EXECUTIVE SUMMARY

The proton processes of all elementary processes are of quasi equal intensities. The neutron processes are "neutral" and driven by the electron processes. The intensities of the electron processes are different, thereby making the difference between the elementary processes. Anti-processes control the elementary balance and the continuity of the elementary evolution in any circumstances. The surplus of the anti-electron processes is the source of the quantum impact of *gravitation*.

The *quantum impulse* is the residual part of the electron process drive of the elementary cycle. The quantum impulse (the *energy quantum*) is of infinite low, never expiring intensity, composing the quantum system of the space-time. Quantum systems are communicating. Quantum impulses are transmitting all quantum impacts without any modification of the quantum information. The speed value of the quantum communication is of infinite large variety and is the main characteristics of the space-time.

The priority of the energy sector today is to find the new energy source.
This book provides clear evidence we do not need to destroy the nature around us in order to produce energy. The key is *gravitation,* the sphere symmetrical expanding acceleration of the *Earth*, the drive of the elementary evolution from the *plasma* to the *Hydrogen* process.

Ref.
S.8

Pyramid is generating energy

The *pyramid* experiment is a "miracle." Not just because it gives new perspectives for energy generation, but because it demonstrates the new look at the world without a word being necessary.

The summarised intensity of the quantum impact of *gravitation* through the basic horizontal surface of the pyramid is higher than its quantum impact leaving the pyramid through its casing, through the other four surfaces.

The quantum speed of the quantum impact of *gravitation* through the casing shall correspond to the quantum speed of the surrounding quantum system. Otherwise, the fluent transition would be broken. The number of the "feeding" quantum impacts of *gravitation* through the base with less territory shall be higher, resulting in internal intensity increase. This intensity difference between the incoming and the leaving impacts results in electron process *blue shift* surplus and conflict within the pyramid.

The surface related absolute balance of the pyramid is in order.
The quantum impact of *gravitation* is anti-electron process *blue shift* impact. The electron process intensity increase of the internals of the pyramid has its impact on the electron processes of the electric wires of capillary size built into the structure of the pyramid as well. And the electron process surplus generating within the wires can be taken out.

Ref.
Annex
8.1
S.8.4

Our inherited pyramids obviously do not have wires inside. Natural granite and limestone are without wires. The replica of the pyramid experiment was built up from a concrete structure with specific mineral composition and with wires inside. The generated potential difference within the replica – with *784 mm* length and *500 mm* height – was about *250-830* mV (depending on the material of the contacting cable-end with the *Earth* surface).

If the generating voltage would depend only on the mass of the pyramid, the supposed voltage potential of a pyramid with *11 m* length and *7 m* height would be *6 kV*. But the measured voltage represents only the potential difference between the two ends of the wire.

It is, without question, that those who built our existing and much admired pyramids knew more than we do right now! The generating increased electron process potential (energy) within the pyramids has been transported by the all over quantum system to a specific "place", a *space-time*, with the quantum speed of the *space-time* equal to the quantum speed, generated by the electron process conflict within the pyramid – with the clear objective of using the intensity/energy of *gravitation*!

Quantum communication is distance-independent, function only of the speed value of quantum communication.

The generating by the pyramid voltage is the most important, but the experiments also prove the energy potential of the *Earth*, its feeding/loading anti-electron process *blue shift* impact. This feeding potential was measured in March-April in Hungary and it was in the range of $(-1) \div (-55)$ mV depending on the weather conditions and the part of the day. This energy potential is the measured voltage between the *Earth* surface and the cable outlet at the top of the pyramid. The measured voltage is representing the feed/load impact of the *Earth* through the base surface of the pyramid – the flow of electron processes <u>from the Earth</u> <u>towards the pyramid</u>.

Ref.
S.6

The acceleration of the *Hydrogen* process is generating energy

Ref.
Table 1
of
S.1

The elementary processes are products of the elementary evolution from the *plasma* to the *Hydrogen* process. The *Hydrogen, Helium, Nitrogen, Carbon, Oxygen, Sulphur, Silicon* and *Calcium* processes are unique ones, as these processes are with proton process dominance. All of them have electron process *blue shift* surplus and these elementary processes are very active in quantum communication.

Ref.
S.1.3

We can say these elementary processes have life-important energy intensity surplus. These are not the ones with the "energy" themselves, rather those which initiate the quantum communication with the others for generating the life-important processes.

These are the elementary processes, capable, in certain mineral compositions, of supporting the rehabilitation of damaged elementary processes, *isotopes*; to significantly shorten their half-life.

Ref.
S.1.4
S.3
Ref.
S.6

All other elementary processes of the Periodic Table and within the minerals of the mantle of the *Earth* are with neutron process dominance.

Neutron process dominant elementary processes have anti-electron process *blue shift* surplus. The surplus is acting and the anti-electron process *blue shift* impact of the surplus is the source of the quantum impact of *gravitation*.

The quantum impact of *gravitation* is formulating within the *Earth* as part of the generating elementary processes of the elementary evolution from the *plasma* to the *Hydrogen* process.

Proton process and neutron process dominant elementary processes behave differently. While proton process dominant elementary processes initiate elementary communication, neutron process dominancy is characterised by internal energy and by feeding/loading the quantum impact of *gravitation.*

Hydrogen process is with electron process *blue shift* drive of infinite low intensity. This is the reason the neutron process of the *Hydrogen process* cannot be measured. The intensity of the collapse of the (d-s-b) quark chain of the neutron process is of infinite low value and the neutron process itself is not measurable within our space-time.
The quantum communication of the natural *Hydrogen* process, however, has been speeded up by the quantum impact of *gravitation* to the quantum speed (the speed of light) on the *Earth* surface. This is the reason the *Hydrogen* process is in volatile gaseous state above the *Earth* surface.

The speeding up of the *Hydrogen* process within an accelerator to $v = i = \lim a\Delta t = c$, the speed value close to the quantum speed on the *Earth,* the intensity of the quantum impact of its electron process increases close to infinite high value.

$$e_e = \frac{dmc^2}{dt_i \varepsilon_H \sqrt{1 - \frac{v^2}{c^2}}} \left(1 - \sqrt{1 - \frac{(c-i)^2}{c^2}}\right)$$

Ref. 6F3

The parameters of the *Hydrogen* process will be changed within the accelerating channel:
   The electron process *blue shift* impact of infinite low intensity of the volatile gaseous state with no conflict with *gravitation* will be of increased to infinite high intensity value (the result of the speeding up). And this increased electron process intensity will be in conflict with the quantum impact of *gravitation.*
The conflict generates electron process *blue shift* surplus and heat.

Ref. S.6.4

The generating electron process surplus might be used without any intermediate steps.
The extremely high heat generation is demonstrated and proven by the acceleration of the *Hydrogen* process for the purpose of collision.
Without the high energy demand for focusing the flow of the *Hydrogen* processes for the purpose of collision, the task, however, is "simple": The external work intensity demand for the acceleration "only" is less than the intensity of the heat generation. The difference is the produced energy for use.
The energy consumption of the acceleration can be optimised by the speed value and the volume of the *Hydrogen* process of the acceleration.

The acceleration of the *Hydrogen* process for the purposes of fusion is not just meaningless but also dangerous. Fusion is a process opposite to the direction of the natural elementary evolution, with all its safety and economic consequences.

The anti-processes control the balance of the elementary processes

Ref. S.3

The neutron process as collapse is driven by the electron process. The intensity value of the collapse shall correspond to the intensity of the anti-neutron process, the expansion within the anti-direction. The elementary process can continue only if the intensity of the anti-proton process – which is the collapse within the anti-direction, driven by the anti-electron process – corresponds to the intensity of the proton process (with the obvious correction by the *entropy*). In other words, the intensity of the anti-proton process determines the intensity of the proton process of the following elementary cycle.

Anti-processes are fulfilling the controlling function of the elementary cycles, keeping the elementary processes on the course of the overall elementary progress.

The quantum speed values of the direct and the anti-processes are equal. The elementary process is one and the same and the quantum membrane of the elementary process operates with equal quantum speed values on the direct and the anti-process sides. The intensities of the electron and the anti-electron processes, however, are different. Their values are reciprocal to each other, but the time system (infinite length) of their operation is the same.

While the world is with an infinite variety of elementary processes, with an infinite variety of intensities and quantum speed values, the basics are identical. The intensities of the quark processes are within the same limits for all elementary processes;

Ref.
S.1.2
1G4
1G7
Ref.
S.2
2A9

    the product of the *square of the quantum speed and the intensity* of the electron process drive is equal for all elementary processes; $\qquad c_x^2 \cdot \varepsilon_x = const$

    the intensity quotient (the *IQ* drive), the quotient of the *square of the quantum speed and the intensity of the anti-electron process* are quasi equal to all elementary processes, (calculating only with $\qquad IQ_- = \dfrac{c_x^2}{\varepsilon_{x-}} = const$

    the decreasing impact of the entropy of the processes):

Elementary processes have different quantum speed values, starting from the infinite high quantum speed value of the *plasma* and finishing with the infinite low quantum speed value of the *Hydrogen* process.

Quark processes represent the proton/neutron, anti-proton/anti-neutron relations.

The neutron process is neutron process, because the (d-s-b)/($\bar{b}$ - $\bar{s}$ - $\bar{d}$) *inflexion* misses its covering (t-c-u)/($\bar{u}$ - $\bar{c}$ - $\bar{t}$) part. The proton process is in the contrary:

Ref.
S.3.3
Diag.
3.5

    the ($\bar{u}$ - $\bar{c}$ - $\bar{t}$)/(t-c-u) *inflexion* is without the ($\bar{b}$ - $\bar{s}$ - $\bar{d}$)/(d-s-b)/ cover.

The other two quark processes in the proton and the neutron processes and also the two anti-quark processes within the anti-neutron and the anti-proton processes are balanced.

The electron and anti-electron processes are the quantum drives of the neutron and the anti-proton processes. The electron process *blue shift* impact or surplus of increased intensity results in the intensity increase of the quantum systems: a quantum membrane.

The anti-electron process *blue shift* surplus of the neutron process dominant elementary processes generates the quantum impact of *gravitation*. The electron process surplus of the proton process dominant elementary processes initiates elementary communication.

Ref.
S.3.6

The intensities of all proton processes *are quasi equal.*

The intensity of the neutron process has been established by the intensity of the electron process. Neutron processes are neutral, are the "passive" ones.

Ref.
S.2.1

The quantum impulse (*quantum*) is the corner stone of the space-time and of the elementary progress

The sequence of the proton-electron-neutron-(anti-neutron)-(anti-electron)-(anti-proton) processes gives the elementary cycle. Elementary cycles develop *entropy.*

There are infinite numbers of elementary events-process cycles in the elementary evolution from the *plasma* state of the *Earth* to the *Hydrogen* process.

In terms of the energy balance there is no perfect cycle. This is not about "nature does not allow" having perfect cycle, rather the fact that "nothing" as such does not exist. There is no way an elementary electron process *blue shift* impact drive could fully expire, could run out to its intensity impact of *zero* value. There is always a remaining infinite low intensity portion. Therefore, the quantum impact of the cover of the proton process is never equal to the quantum impact received by the neutron process. (Anti-processes control the direct process therefore, they do not act independently.)

Quantum impulse is generating within the elementary cycle at the *inflexions* between the neutron/anti-neutron and the anti-proton/proton processes.

There is always a residual portion, an infinite small intensity impulse within the electron process *blue shift* drive, which cannot be utilised

$$\lim \Delta_{quantum} = (1-\zeta)\frac{dmc^2}{dt_i \varepsilon_e}\left(1-\sqrt{1-\frac{(c-i)^2}{c^2}}\right) = 0;$$

where $\lim_{0 \to 1} \zeta = 1$

Ref. 2D5

There are two consequences here:
- an entropy value in the *driven* side, and
- a quantum impulse value (energy quantum) on the *driving* side.

The quantum impulse of the elementary processes is not equal to the anti-electron process *blue shift* impact of the neutron process dominant elementary processes, feeding *gravitation*. These two categories are clearly different and are acting in parallel. The quantum impulse is establishing the quantum system of reference.

One hundred years after *Max Planck* and *Einstein,* it must be emphasised that quantum impacts and signals are not about flying photons. Quantum impacts and signals are transported by *quantum impulses* (energy quantum).

Max Planck said in a conference at Florence, Italy, in 1944:

"As a man who has devoted his whole life to the most clear headed science, to the study of matter, I can tell you as a result of my research about atoms this much: There is no matter as such. All matter originates and exists only by virtue of a force which brings the particle of an atom to vibration and holds this most minute solar system of the atom together.
We must assume behind this force the existence of a conscious and intelligent mind. This mind is the matrix of all matter."

Ref.*

"There is no matter as such" is *Planck's* message. All what we measure are processes with the infinite high variety of the intensities and the quantum speed values.

[Mass and energy as categories have been used in the book as well, in a couple of cases only, for better understanding and for easier explanations. Mass as such is the interpretation of the measured weight, the intensity impact of the process at the *inflexion* of the event. The case with the energy is easier, since it can be understood as the intensity of the events.]

Ref. S.5

Space-time matrix

Ref. S.2

Space-time is about quantum impulses. *Space-time* has been built up from *quantum impulses* – events of infinite low intensity – composing altogether an aggregate system. *Space-time* – the system of quantum impulses – is transporting/transferring quantum impacts and signals by the speed of the quantum impacts and signals.

Space-time is not about distance. *Space-time* is about quantum speed and intensity.

The transportation of quantum impacts and signals is the quantum communication itself.
Quantum impulses (the energy quantum) do not take quantum signals;
    do not add anything to the impacting and transported quantum signals;
    they are impulses themselves with *blue shift* impact of infinite low intensity.

Ref.
5C4

As being the smallest possible energy/intensity quantum, quantum impulses cannot disappear, as "nothing" as such does not exist; they do transfer quantum signals in any direction; *in this way they represent the highest power of the universe*: whatever the impacting signal is, the energy quantum remain as they were before – infinitely small, never expiring quantum impulses without change.
Whatever elementary process would be the origin of the quantum impulse, there is no difference in the acting intensity. They have been born, are and remain of infinite low intensity.

Ref.
S.
5.2.3

Distance, as such, for the measurement of space has "only" practical importance.
The definition of the space by length dimensions is the conventional way of measuring it.
Space-time is about quantum impacts with no limitation of intensity and time!

Ref.
S.5.2

Thus, the definition of *space-time* is given by events (again)!
The same way as it is for *time* – just the event for counting the *time* is more obvious and, therefore, is more acceptable than for *space*.
The real measurement of *space-time* in any direction is the number of the impacted quantum impulses. The higher the quantum speed is, the "bigger" is the space-time, the higher is the number of the impacted quantum. The higher number of the impacted energy quantum means higher intensity. In a *space-time* of increased intensity the time flows *more slowly*.
With the increase of the quantum speed, the space (the world) is "opening".

The *space-time matrix* is the complexity of quantum systems with the infinite large variety of the quantum speed. *Space* and *time* are containing each other. Elementary processes generate their own space-time as the speed values of their quantum communication are different. The *space-time matrix* contains the space-time of all elementary processes built up on each other.
With this *space-time* definition, the *Einstein* time formula, with *v*, the speed value of events and *Planck* quantum views, are hand in hand together. This *space-time* definition makes it possible for referring equally and in parallel to *time* and *space*, but, most importantly, for understanding the meaning of the *space-time*.

Space-times exist in parallel: different *space-times* means having quantum impacts in the quantum system by the variety of quantum speed values.
We live within our *space-time* on the *Earth* surface with our acting quantum speed of $c = 299,792$ km/sec, the speed of light. We sense our events in line with our time flow.
These "our events" on the surface of the *Earth* might have different durations within other space-times of different quantum speed values. For us, however, only our events with the quantum speed of light on the *Earth* surface can be "registered". We cannot receive any other information, just the one provided by our space-time and time system.

Our space-time on the *Earth* surface is generated by the anti-electron process *blue shift* impact of the *plasma* and the evolution of the elementary processes = *gravitation.*

Because *space-times* are about the number of the impacted quantum, quantum impulses can be impacted by the variety of quantum speed values.

As quantum impulses have not just been generated by elementary processes of different quantum speed and intensity values, but might also be impacted by the electron process *blue shift* impacts of the same variety of elementary processes, *space-times* exist and operate in parallel!

The parallel operation means: the quantum systems are impacted by the infinite large variety of elementary impacts.

The parallel existence of space-times raises two effects: *communication* and *conflict.*

Elementary *communication* is about establishing integrated elementary structures by the use of the existing quantum potential of the elementary cycles.

Conflict means impacting the same energy quantum by two or more signals, where the conflict itself is generating between the impacting signals and never with the quantum impulse. Quantum impacts, propagating in *space-times,* lose on their intensity value, because of the developing conflicts during the transfer.

The most difficult subject for the conventional mind to accept is that the *space-times* are *distance- independent.* There is no difference where the quantum signals of equal quantum speed are acting – they are equally part of the same *space-time.*

There are no such categories within the *space-time matrix* as "far" or "close", "short" or "long distance". While quantum signals, in conventional terms, might be impacting the quantum system "close" to, or "far" from each other, the generating *space-time* depends only on the speed value of the quantum communication.

Quantum signals of less quantum speed than the operating quantum speed of our *space-time* cannot impact us. (As a quantum effect of less intensity cannot impact a system with higher intensity.) Higher quantum speed signals at the same time as generated by pyramids though cannot be registered. Higher quantum signals only have measurable impact if they are not local ones (as pyramids are) and are causing conflict as the *sunshine* does.

The *space-time matrix* is the world itself, with *space-times* of infinite large variety of quantum speed values and intensities.

Time is established by the intensity of events. The *time* is the relation between the event and the quantum system, the indicator of the intensity of the quantum impact.

---

    *   Das Wesen der Materie; speech at Florence, Italy (1944) (from Archiv zur Geschichte der Max-Planck-Gesellschaft, Abt. Va, Rep. 11 Planck, Nr. 1797) – taken from Wikipedia "Max Planck".

Table of content

1   The *IQ* of the elementary evolution   1

  1.1   *Intensity Quotient, IQ* of elements – the key of quantum
        communication   5
  1.2   The quantum speed and the process/anti-process relation   10
  1.3   The quantum impact of *gravitation*   13
  1.4   *Hydrogen* process and *gravitation*   14
  1.5   The conflict with the intensity of the *gravitation* on the *Earth* surface 15
  1.6   From *plasma* state to the *Hydrogen* process   17
  1.7   Space-time has been established by the speed of quantum
        communication   19

2   The *quantum impulse (energy quantum)*   21

  2.1   The quantum impulse   23
  2.2   The harmony of the elementary process   28
  2.3   The quantum system of the *Earth*   32
  2.4   Space-times, difference in time counts   33
  2.4.1 The value of the speed of the quantum communication on the *Earth*   35
        surface has been established by *gravitation*
  2.4.2 The impact of the *Earth* space-time on elementary processes   35
  2.4.3 Time impact of the quantum membrane of increased intensity   37
  2.4.4 The space-time of the *Earth* surface   37
  2.5   The harmony with *h*, the *Planck* constant and the *Planck* formula   38

3   Process/anti-process relations   39

  3.1   Proton process dominance   44
  3.2   Neutron process dominance   46
  3.3   Elementary communication   47
  3.3.1 The intensity of the electron process in elementary communication   48
  3.4   *Hydrogen* and the anti-process   51
  3.5   Anti-processes and isotopes   52
  3.5.1 Fusion as such is not a practical option   54
  3.6   Important conclusion: the intensities of all proton processes are equal   55

4	**Quantum communication in practice**	56
4.1	Mass/energy intensities to be utilised	58
4.2	The way of the formulation of the surplus	59
4.3	Quantum communication of elementary processes	60
4.4	Anti-processes	62
5	**Space-time matrix**	65
5.1	Elementary space-times are embedded within each other and formulating *matrix*	68
5.2	Communication of space-times	69
5.2.1	The space-time of the *Earth* surface	72
5.2.2	*Inflexions* of events	73
5.2.3	Space-times are "distance-independent"	73
5.3	The duration and the intensity of events having been influenced by motion	75
5.4	Infinite low intensity is turning into infinite high value	75
6	**Energy generation by the electron process *blue shift* conflict of the *Hydrogen* process and *gravitation***	79
6.1	*Plasma*	79
6.2	After *Plasma* approaching finally to the *Hydrogen* process	83
6.3	Acceleration of the *Hydrogen* process	86
6.4	Utilisation of the benefit	87
6.5	The work value of the speeding up and the benefit of the conflict with *gravitation*	89
6.6	*Hydrogen* acceleration and space-time	92
7	***Isotopes* and *isotope* rehabilitation**	95
7.1	Beta radiation	95
7.1.1	*Beta(-)* radiation means the elementary process is with missing electron process *blue shift* impact.	95
7.1.2	*Beta(+)* radiation means damage in the electron process as well, just in the opposite direction.	98
7.2	Alpha radiation	100
7.3	Gamma radiation	101
7.4	X-ray	105
7.5	Neutron radiation	105
7.6	Rehabilitation of isotopes by external support	106
7.6.1	For proving the statement about the rehabilitation isotopes a decontamination mix was prepared and presented in details within Annex 7.1.	110

8   **Quantum impacts of *pyramids* - the power**   111

8.1   *IQ* and the intensity relations   118

8.2   Intensity distribution   120

8.3   *Pyramid* is generating energy
Pyramid is a potential power plant
Results and the conclusion of the measurements with the *pyramid* replica   123

8.4   The quantum impact of *Earth gravitation* is transported to a different   133
space-time

8.5   *Sunshine*, the quantum impact   133

    Annexes   135

1.1   The *Genesis*   137

2.1   Space-time impact for human relations   140

3.1   Additional thoughts about processes and anti-processes   141

4.1   Quantum communication in practice   146

5.1   Rotating disc experiment with switching off and on light impacts   154

6.1   Hartmann and other impacts   158

6.2   Process based assessment of the "walkers" of oil bath   162

7.1   Rehabilitation of KOH by decontamination mix   167

7.2   Rehabilitation of Cs-137 by decontamination mix   170

7.3   Rehabilitation of RBMK graphite   174

8.1   The pyramid replica experiment   179

8.2   Building structures   192

8.3   The cooling of the elementary processes of the *Earth* by *gravitation*   197

1
The *IQ* of the elementary evolution

1. The neutron process of the *plasma* is collapse of infinite high intensity. The collapse is driven by electron process *blue shift* impact of infinite high intensity.
2. The anti-neutron process, the result of the *inflexion* of the neutron process, is expansion of infinite high intensity.
3. The anti-neutron process generates the anti-electron process, the *blue shift* drive for collapsing the anti-proton process.
4. The proton process, the result of the *inflexion* of the anti-proton process, generates electron process *blue shift* impact and drives the neutron process.
5. The *blue shift* impacts of either the electron or the anti-electron processes are in surplus.

The *plasma* state is overall elementary neutron collapse of infinite high intensity.

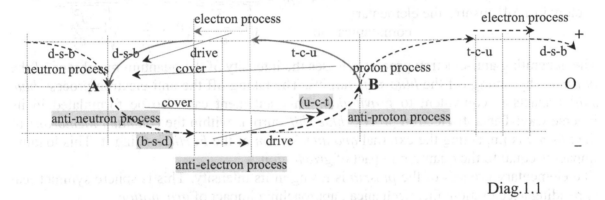

Diag.1.1

Diag. 1.1 demonstrates the case.

There is only one cycle of the (t-c-u)–(d-s-b) process chain presented with their anti-processes on the diagram. *Inflexion* points A and B are the same, as space-times are about intensities rather than space dimension coordinates (as will be proven later in this section). The (d-s-b) and the (b-s-d), the (u-c-t) and the (t-c-u) *inflexions* happen as parts of the elementary cycle.

The intensity of the neutron collapse at the *plasma* is $\varepsilon_n = \infty$ and whatever the intensity of the proton process is, the intensity coefficient of the electron process drive of the collapse is $\lim \varepsilon_{pl} = 0$ and the energy intensity of the drive is $\lim e_{pl} = \infty$.

$$\varepsilon_e = \frac{\varepsilon_p}{\varepsilon_n}\sqrt{1-\frac{(c-i)^2}{c^2}}; \qquad \text{and} \qquad \lim e_{pl} = \lim \frac{dmc_{pl}^2}{dt_i \varepsilon_{pl}}\left(1-\sqrt{1-\frac{(c_{pl}-i_{pl})^2}{c_{pl}^2}}\right)=0 \qquad \text{1A1}$$

The *Hydrogen* process is the other end.

The intensity of the neutron process is $\lim \varepsilon_n = 0$, in this way, the intensity coefficient of the electron process is $\lim \varepsilon_H = \infty$ and the electron process *blue shift* drive of the neutron collapse is $\lim e_H = 0$.

$$e_H = \frac{dmc_H^2}{dt_i \varepsilon_H}\left(1-\sqrt{1-\frac{(c_H-i_H)^2}{c_H^2}}\right) \qquad \text{1A2}$$

Ref. S.5.4 The *plasma* state is the result of the infinite high quantum drive with infinite high quantum speed and infinite high intensity ($\lim \varepsilon_{pl} = 0$) of the electron process *blue shift* impact.

The *Hydrogen* process is, on the contrary, the result of the infinite low quantum drive of the collapse and the infinite low quantum speed of the elementary process.

The infinite high intensity of the neutron collapse of the *plasma* results in electron process *blue shift* surplus and conflict on the anti-process side. The reason is that the intensity of the collapse of the anti-proton process at the *inflexion* should correspond to the intensity of the expansion of the proton process on the other, direct side. The proton process generates electron process *blue shift* impact without surplus, but of increased intensity.

The quantum speed values are equal on the direct and the anti-process sides, Z_{pl} the event concentration (the reciprocal value to ε_{pl}) is of infinite high value.

$$Z_{pl} = \frac{\varepsilon_n}{\varepsilon_p} = \frac{dt_p}{dt_n} \quad \text{where} \quad \lim \Delta t_n = 0$$

1A3

The time system of the electron processes remains practically unchanged, independently of the intensities of the processes and the values of the speed of quantum communication.

This quasi equal dt_i time system, shown below in 1A4, ensures the elementary communication:

1A4

$$dt_i = \frac{dt_o}{\sqrt{1 - \dfrac{i^2}{c^2}}} \cong \frac{dt_o}{\sqrt{1 - \dfrac{i_{pl}^2}{c_{pl}^2}}} \cong \frac{dt_o}{\sqrt{1 - \dfrac{i_x^2}{c_x^2}}} \cdots$$

The generating anti-electron process releases the intensity of the quantum membrane of the *plasma* – by giving off the *blue shift* surplus. The taking off the anti-electron process *blue shift* impacts is equivalent to *gravitation*. This statement can also be formulated in its inverse way: The anti-electron process *blue shift* surplus within the *Quantum Membrane* of the *plasma* is impacting the external *quantum system* of the *Earth*, loading it. This loading impact is equal to the quantum impact of *gravitation*.

The elementary process of the *plasma* is losing in its intensity. This is sphere symmetrical expanding acceleration, the mechanical, approaching impact of *gravitation*.

With the load of the external *quantum system*, the intensity of the *blue shift* surplus (and conflict) on the anti-process side of the *plasma* state is decreasing, as part of the *blue shift* impact is taken away:

As a consequence, the resulting and acting *blue shift* impact within the elementary processes is: $[g < 1]$

$$e_{n-r} = (1-g)\frac{dmc_{pl}^2}{dt_i \varepsilon_{pl}}\left(1 - \sqrt{1 - \frac{(c_{pl} - i_{pl})^2}{c_{pl}^2}}\right)$$

1B1

Elements are formulating.

The effect, in accordance with 1B1, would theoretically result in less electron process *blue shift* impact in numbers than it was at the earlier *plasma* state, with infinite high speed of quantum communication and of infinite high intensity of the electron *process*.

Electron process endpoints, however, are the starting points of the neutron processes. The intensity of the *blue shift* impact is the one which has been reduced – the number of the electron processes remains unchanged. The acting drive, the consequence of *gravitation*, corresponds therefore to the same number of electron processes, but with less intensity.

1B2

$$e_g = (1-g)\frac{dmc_{pl}^2}{dt_i \varepsilon_{pl}}\left(1 - \sqrt{1 - \frac{(c_{pl} - i_{pl})^2}{c_{pl}^2}}\right) = \frac{dmc_x^2}{dt_i \varepsilon_x}\left(1 - \sqrt{1 - \frac{(c_x - i_x)^2}{c_x^2}}\right); \quad \begin{array}{l} \text{with } c_x < c_p \\ \text{and } \varepsilon_x > \varepsilon_{pl} \end{array}$$

The reduction of the *blue shift* impact relates to the quantum speed in direct quadratic, to the intensities of the electron processes in linear but inverse proportions.

$$1 - g = \frac{\varepsilon_{pl}}{c_{pl}^2} \frac{c_x^2}{\varepsilon_x}$$

1B3

The impact in the brackets in 1B2 can be taken as quasi equal.

If $g = 0 \rightarrow \dfrac{c_{pl}^2}{\varepsilon_{pl}} = \dfrac{c_x^2}{\varepsilon_x}$; no cooling by *gravitation*.	If $\lim g = 1 \rightarrow \lim \dfrac{c_x^2}{\varepsilon_x} = 0$; taken all.

1B4

With reference to 1B3 and 1B4,
the more *blue shift* impact is taken by *g gravitation* $\Rightarrow$ the less is the remaining $(1 - g)$ portion. The more is the value of the ε_x coefficient,

➢ the less the energy/mass intensity of the electron process *blue shift* impact is, the less is the intensity of the neutron collapse;
➢ the less is the value of c_x the speed of quantum communication.

As a consequence, with the "expansion" of the *plasma* – cooling by *gravitation*:
➢ the intensity of the electron process in the formulating elementary processes within the remains of the *plasma* (*Earth* core) are decreasing towards the *Earth* surface; as energy intensities (e) are reversely proportional to the intensity coefficients (ε); and
➢ the speed value of the quantum communication is decreasing.

[The expansion itself is no other than *blue shift* impact.]

Earth plasma as *blue shift* conflict of infinite high intensity is the origin of *gravitation*. The elementary processes within the core and within the mantle of the *Earth* are the remains of the *plasma* and have also anti-electron process *blue shift* surplus for feeding/loading *gravitation*. The electron processes within the establishing core/mantle of the *Earth* structure, cooled by *gravitation,* are communicating (reference to 1A4) – as having quasi identical time systems.

Ref. 1A4

The *plasma* and all other elementary processes of the core contribute to the quantum impact of *gravitation* by their anti-electron process *blue shift* surplus through the mantle of the *Earth*. The conflict generates heat and develops quantum membrane above the surface. The more the *blue shift* conflict is, the more is the heat.

Balanced transition means that

if the parameters of the *plasma* are $\quad\quad\quad\quad c_{pl}$ and ε_{pl} ;

and the external *Quantum Membrane* of the *Earth* is of $\quad c_{qm}$ and ε_{qm} ;

the intermediate core structures between the two have: $\quad c_{pl} > c_x > c_{qm}$ and

$$\varepsilon_{pl} < \varepsilon_x < \varepsilon_{qm};$$

The intensity of the electron process *blue shift* impact within the core, distancing from the *plasma* towards the external *Quantum Membrane,* is less and less. For corresponding to the same cooling (*gravitation*) energy, the number of elementary processes should be different.

$$k \frac{dmc_{xa}^2}{dt_i \varepsilon_{xa}} \left(1 - \sqrt{1 - \frac{(c_{xa} - i_{xa})^2}{c_{xa}^2}} \right) = r \frac{dmc_{xb}^2}{dt_i \varepsilon_{xb}} \left(1 - \sqrt{1 - \frac{(c_{xb} - i_{xb})^2}{c_{xb}^2}} \right) = \ldots =$$

$$= s \frac{dmc_{xy}^2}{dt_i \varepsilon_{xy}} \left(1 - \sqrt{1 - \frac{(c_{xy} - i_{xy})^2}{c_{xy}^2}} \right)$$

1B5

With the increase of ε_x, c_x the quantum speed is decreasing. $k < r < ... < s$ means the proportions of the formulating elementary processes towards the surface are increasing.

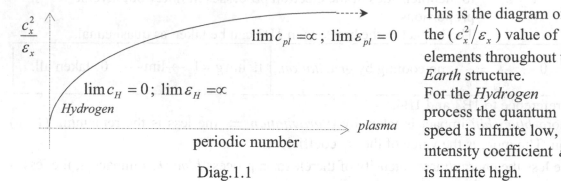

This is the diagram of the (c_x^2 / ε_x) value of elements throughout the *Earth* structure.
For the *Hydrogen* process the quantum speed is infinite low, the intensity coefficient ε_H is infinite high.

Diag. 1.1

$\lim c_{pl} = \infty$; $\lim \varepsilon_{pl} = 0$

$\lim c_H = 0$; $\lim \varepsilon_H = \infty$

Hydrogen

plasma

periodic number

Diag.1.1

The increasing electron process intensity within the core towards the *plasma* results in increased neutron process intensity indeed: The increased intensity of the electron process *blue shift* drive results in increased speed of quantum communication. The cooling of *plasma* by *gravitation* is resulting in expansion, at the same time in the shrinking of the *plasma* state, the generation of a certain crust cover above and around the *plasma* with elementary process structures of minerals and soil.
The quantum speed values and the intensities of the electron process *blue shift* impact of elementary processes of minerals and soil are decreasing towards the *Earth* surface.

Alongside with the cooling effect of *gravitation* (the load from the anti-electron process *blue shift* impact and surplus), elementary processes start formulating with proton and neutron processes of certain intensity values and relations within the crust of the *Earth*. The gradient of the load (of *gravitation*) towards the *Earth* surface is decreasing.
The intensities of the neutron processes of elementary processes away from *plasma* have also similar decreasing tendency: the intensity of the electron process drive is getting less and less. The event concentration of the process is decreasing – equivalent to *sphere symmetrical expansion*.
[Elementary communication is about elementary processes of three consecutive cycles. Electron, proton and neutron processes of the same cycle cannot be the drive, the driven and the coverage at the same time.]

1C1

The equality of the *Strong Interrelation* is always valid:

$$\frac{dmc_x^2}{dt_p \varepsilon_p}\left(1 - \sqrt{1 - \frac{i^2}{c_x^2}}\right) = \xi \frac{dmc_x^2}{dt_n \varepsilon_n}\sqrt{1 - \frac{(c_x - i_x)^2}{c_x^2}}\left(1 - \sqrt{1 - \frac{i^2}{c_x^2}}\right)$$

ξ demonstrates the developing in each elementary cycle the value of quantum entropy;

1C2

But in intensity terms the difference can be significant:

$$\frac{dmc_x^2}{dt_p}\left(1 - \sqrt{1 - \frac{i^2}{c_x^2}}\right) \neq \frac{dmc_x^2}{dt_n}\sqrt{1 - \frac{(c_x - i_x)^2}{c_x^2}}\left(1 - \sqrt{1 - \frac{i^2}{c_x^2}}\right)$$

As the result of *gravitation*

➤ the intensity coefficient of the electron process is approaching $\varepsilon_x = 1$ from $\lim \varepsilon_{pl} = 0$;

➤ ε_p the intensity of the proton process is increasing and approaching ε_n, the intensity of the neutron process;

➤ the speed of quantum communication is approaching the value, measured on the *Earth* surface (from $\lim c_{pl} = \infty$).

Gravitation is the result of the *blue shift* the impact of the anti-electron processes.

The following intensity decrease of the electron process (meaning: $\varepsilon_x > 1$ - the increase of the electron process intensity coefficient) is equivalent to electron process *blue shift* surplus and proton process dominance. The electron process *blue shift* impact is acting at this level at "low" speed value of quantum communication, generating liquid and gaseous state rather than *plasma*. Both are of certain conflicting character.

There are *eight* elementary processes within the periodic table with proton process dominance and electron process *blue shift* surplus: the *Hydrogen, Helium, Oxygen, Nitrogen, Carbon, Silicon, Sulphur* and *Calcium* processes, with the intensities of the neutron processes less than the intensities of the proton processes. Their appearance (gaseous, powder and stable) depends on the balance of their proton and neutron processes.

In conclusion, it can be stated that the cooling process by *gravitation* starts and also ends by conflicting electron process *blue shift*.

But the conflict
➤ at the start is on the anti-electron process side of the *plasma*, with infinite speed value and infinite high intensity of quantum communication,
➤ at the end it is on the electron process side of the *Hydrogen* process, an elementary status of infinite low values of quantum speed and of the *blue shift* surplus.

1.1
Intensity Quotient IQ of elements – the key of quantum communication

S.
1.1
Ref.

$$\frac{c_x^2}{\varepsilon_x}$$

gives (with reference to 1B4) the real intensity drive of the electron process;

c_x is the speed value of all, the proton, the neutron and the electron processes;

ε_x is the intensity coefficient of the electron process.

1B4
1D1

For calculating the *IQ* value for each elementary process, we have to find the speed value of the quantum communication for each of the elementary processes.

$$\frac{\dfrac{dmc_x^2}{dt_i}}{\dfrac{dt_{nx}}{dt_{px}}} = \frac{dmc_x^2}{dt_i dt_{px}} dt_{nx} = \frac{dmc_x^2}{dt_i}\varepsilon_x = const$$

The intensity of the unified neutron collapse relative to the intensity of the proton process is one and the same for all elementary processes! This ensures the proton process cover and electron process drive in any elementary communication.

1D2

All conditions of the elementary quantum communication are ensured:
(1) the time system of the electron processes is one and the same for all elementary processes;
(2) the neutron process can be driven by the electron process of any elementary processes. The collapse, (the neutron process), is "*neutral*" indeed.
(3) the proton process intensity of the unified neutron process intensity drive is one and the same for all elementary processes.

For having identical unified proton process intensity values: $c_x^2 \cdot \varepsilon_x = c_y^2 \cdot \varepsilon_y = ... = c_z^2 \cdot \varepsilon_z$

1D3

$\dfrac{dm}{dt_i}$ value has been left out from 1D3, as this part of the equation is equal for all elementary processes. The square of the quantum speed value is compensating the change. In the case of higher ε_x intensity coefficient value, c_x becomes less; in less ε_x value, c_x becomes more.

1D4

The quantum speed values are the same and equal in the anti-, and the direct processes. The intensity of the unified anti-proton collapse is with the correction of the generating intensity coefficient. The *blue shift* surplus is given off for *gravitation*:

1D4
$$\frac{dmc_x^2}{dt_i}\left(\varepsilon_{-x}-\varepsilon_g\right)=\frac{dmc_x^2}{dt_i dt_{-nx}}\left(dt_{-px}-\Delta_g\right)=\frac{dmc_x^2}{dt_i}\varepsilon_{-xm}; \qquad \varepsilon_{-xm}\text{ - modified intensity}$$

1D6

as $dt_{nx}=\left|dt_{-nx}\right|$ and $dt_{px}=\left|dt_{-px}\right|$ and $\dfrac{dt_{nx}}{dt_{px}}\neq\dfrac{dt_{-px}}{dt_{-nx}}$; meaning: $\varepsilon_x\neq\varepsilon_{-x}$ the relation in 1D2 would give c_{-x} different quantum speed	$\dfrac{\dfrac{dmc_{-x}^2}{dt_i}}{\dfrac{dt_{-px}}{dt_{-nx}}}=\dfrac{dmc_{-x}^2}{dt_i dt_{-nx}}dt_{-px}=\dfrac{dmc_{-x}^2}{dt_i}\varepsilon_{-x}$

The quantum speed should obviously be one and the same: $c_x=c_{-x}$, but the intensity – on account of the load for *gravitation* – covers the difference.

1D7
1D8

$\dfrac{dmc_x^2}{dt_i \varepsilon_{-xm}}\left(1-\sqrt{1-\dfrac{(c_x-i_x)^2}{c_x^2}}\right)$;	is the drive of the anti-proton process	$\dfrac{dmc_x^2}{dt_i \varepsilon_g}\left(1-\sqrt{1-\dfrac{(c_x-i_x)^2}{c_x^2}}\right)$	is the load for *gravitation*

1D9   The unified intensity relation for the direct and anti-processes: $c_x^2\cdot\varepsilon_x\neq c_x^2\cdot\varepsilon_{-xm}$

1D10   The value of $c_x^2\cdot\varepsilon_x$ for all elementary processes are equal and represent the *unified* energy intensity value of the electron process.

Once the quantum speed value of the *Carbon* process is taken for $c_C=300,000$ km/sec on the surface of the *Earth* and ε_x the electron process intensities of all elementary processes are known, the speed values of quantum communication of all other elementary processes can be calculated.

1D11
$$\frac{c_C^2}{Z_C}=\frac{c_x^2}{Z_x}; \quad\text{and}\quad c_x=c_C\sqrt{\frac{Z_x}{Z_C}}; \quad\text{or}\quad c_x^2\cdot\varepsilon_x=c_C^2\cdot\varepsilon_C \quad\text{and}\quad c_x=c_C\sqrt{\frac{\varepsilon_C}{\varepsilon_x}};$$

(While the quantum speed value on the surface of the *Earth* is $c_{Earth}=299,792$ km/sec, the speed of quantum communication of the *Carbon* process is taken deliberately for 300,000 km/sec.)

As an example, the calculation of the quantum speed of the *Aluminium* process is given here below. The intensity coefficient of the *Carbon* process is $\varepsilon_C=1.01338$ and the intensity coefficient of the *Aluminium* process is: $\varepsilon_x=0.94697$.

The corresponding quantum speed of the *Aluminium* process will be:

1D12
$$c_{Al}=c_C\sqrt{\frac{1.01338}{0.94697}}; \quad\text{and}\quad c_C=300,000 \text{ km/sec}$$

Table 1.1 contains the quantum speed and *IQ* values for all elementary processes/minerals. Elementary processes vary

> Elementary processes with their natural speed value of quantum communication less than the quantum speed on the surface of *Earth* ($c_{Earth}=299,792$ km/sec) are in volatile gaseous status. The speed values of quantum communication of *He, O* and *Ni* processes have been calculated by $c_{Earth}=299,792$ km/sec, all others, by $c_C=300,000$ km/sec.

The conflict, generated by the natural *blue shift* surplus of these elementary processes (*H, He, Ni, O*) is strengthened by the acting quantum speed of *gravitation*.

> Elementary processes in their mineral status with higher quantum speed than the quantum speed of *gravitation* and becoming molten on the *Earth* surface, lose on their natural quantum speed value. Melting is *blue shift* conflict of infinite high intensity, which destroys the elementary structure. *Gravitation* establishes the elementary communication again, but the newly formulating quantum speed value will be corresponding to the quantum speed of *gravitation* on the *Earth* surface.

> The melting-out of elementary processes from their minerals status on the *Earth* surface and the following solidification causes additional elementary conflict: Elements have their own ε_x electron process intensity value, which determines the intensity relation of the proton and neutron processes.

Once the element is melted and out of its mineral status, the relating quantum speed is corresponding to c on the *Earth* surface, less than that for the element in its mineral status:

$$\frac{c^2}{\varepsilon_x} < \frac{c_x^2}{\varepsilon_x} \text{ as } c < c_x \qquad \text{1E1}$$

With less quantum speed value (after the melting) the elementary process is in *blue shift* conflict: For reaching the *inflexion* point it needs conflicting internal quantum membrane in order to increase the effect of the decreased quantum speed. Elementary processes with dominant neutron process intensity may, therefore, have *blue shift* conflict on the *Earth* surface! - The best example is *Mercury*, as a side product of heavy metallurgy – liquid, while in fact the element has significant *blue shift* deficit.

But all heavy elements such as *Lead, Gold, Platinum, Uranium* and others become softened after melting out from their mineral status – even having electron process *blue shift* deficit and relatively high speed value of quantum communication.

> The reactions of elementary processes with less original quantum speed value than the quantum speed of *gravitation* $c > c_x$ is the opposite: The cooling by *gravitation* results in increased speed and conflict and they become gaseous.

The speed of quantum communication is an important component.

Oxygen process has natural quantum speed less than the quantum speed of the *Carbon* process. *Hydrogen* process has even less, in fact, infinite small speed value.

The electron process *blue shift* drive of the *Hydrogen* process, having been of infinite low intensity, means that the electron process *blue shift* impact of other elementary processes may drive the quasi not-driven neutrons of the *Hydrogen* process. This, however, is not about "*Hydrogen* process is driven by other elementary process". The driving elementary process is managing its own process, using the available neutrons, quasi not driven within the elementary structure of the *Hydrogen* process.

Water and *Hydrocarbons* are the best examples. The available *blue shift* surplus of the *Oxygen* and the *Carbon* processes is fully used by driving the neutrons available (within the *Hydrogen* process). The quasi never-used electron process *blue shift* surplus of the *Hydrogen* process makes both the *water* and the *hydrocarbons* liquid.

Fire means *blue shift* conflict of infinite high intensity on the surface of the *Earth*.

The speed of the conflict is equal to the speed of the quantum communication on the surface of the *Earth*.

Water, as the result of the elementary communication of the *Oxygen* and the *Hydrogen* processes, has less electron process *blue shift* conflict and quantum speed value than that on the *Earth* surface. The *Hydrogen* process is taking not just the electron process *blue shift* drive from the fire, but also reduces its quantum speed value. In this way, water takes off the *blue shift* conflict and the conflicting speed and extinguishes the fire!

Oxygen process is with electron process *blue shift* surplus. *Oxygen* process is in gaseous state and feeding fire. The gaseous state is the result of *gravitation* and its acting speed of quantum communication becomes equal to c on the surface of the *Earth*.

Hydrocarbons feed fire, as the quantum speed of the *Carbon* process is higher than the quantum speed on the *Earth* surface. The difference comparing with *water* is that while the natural *Oxygen* process keeps the quantum speed of the water at lower level than the speed value on the *Earth* surface, *Carbon* keeps it at higher level. Quantum communication is increasing the conflict.

Water and *Hydrocarbons* in liquid state both have certain *blue shift* surplus.

IQ values of minerals are given in Table 1.1 here below:

Element	PN	Measured atomic weight	$Z_x^{max} = \dfrac{1}{\varepsilon_x^{max}}$	ε_x	c_x	$IQ = \dfrac{c_x^2}{\varepsilon_x}$
Hydrogen	1	1.00790	0.000081	12345.67	2718*	5.984E+02
Helium	2	4.00260	0.9863	1.01389	299716*	8.872E+10
Lithium	3	6.94000	1.2961	0.77155	343815	1.532E+11
Beryllium	4	9.01200	1.2362	0.80893	335777	1.394E+11
Boron	5	10.81000	1.1458	0.87275	323266	1.197E+11
Carbon	6	12.01100	0.9868	1.01338	*300000*	8.881E+10
Nitrogen	7	14.00670	0.986	1.01420	299670*	8.808E+10
Oxygen	8	15.99900	0.9849	1.01533	299503*	8.847E+10
Fluorine	9	18.99840	1.0951	0.91316	316033	1.094E+11
Neon	10	20.17000	1.0019	0.99810	302286	9.155E+10
Sodium (Na)	11	22.98900	1.0743	0.93084	313018	1.053E+11
Magnesium	12	24.30500	1.0102	0.98990	303536	9.307E+10
Aluminium	13	26.98150	1.056	0.94697	310340	1.017E+11
Silicon	14	28.08550	0.9911	1.00898	*300652*	8.959E+10
Phosphorus	15	30.97370	1.0495	0.95283	309348	1.004E+11
Sulphur	16	32.06000	0.9887	1.01143	*300288*	8.915E+10
Chlorine	17	35.45300	1.0699	0.93467	312376	1.044E+11
Argon	18	39.94800	1.2028	0.83139	331210	1.319E+11
Potassium (K)	19	39.09800	1.0424	0.95932	308335	9.910E+10
Calcium	20	40.08000	0.989	1.01112	*300334*	8.921E+10
Scandium	21	44.95590	1.1248	0.88905	320290	1.154E+11
Titanium	22	47.90000	1.161	0.86133	325403	1.229E+11
Vanadium	23	50.94150	1.1983	0.83452	330589	1.310E+11
Chromium	24	51.99600	1.1503	0.86934	323901	1.207E+11
Manganese	25	54.93800	1.1811	0.84667	310340	1.138E+11
Iron (Fe)	26	55.84700	1.1319	0.88347	321300	1.169E+11
Cobalt	27	58.93320	1.1664	0.85734	326159	1.241E+11
Nickel	28	58.71000	1.0811	0.92498	314007	1.066E+11
Cuprum	29	63.54000	1.1746	0.85135	327304	1.258E+11
Zinc	30	65.38000	1.1631	0.85977	325698	1.234E+11
Gallium	31	69.73500	1.2327	0.81123	335301	1.386E+11
Germanium	32	72.59000	1.2515	0.79904	337848	1.428E+11
Arsenic	33	74.92160	1.2534	0.79783	338105	1.433E+11

Selenium	34	78.96000	1.305	0.76628	344994	1.553E+11
Bromine	35	79.90400	1.2659	0.78995	339786	1.462E+11
Krypton	36	83.80000	1.3104	0.76313	345707	1.566E+11
Rubidium	37	85.46580	1.2927	0.77357	340364	1.498E+11
Strontium	38	87.62000	1.2886	0.77604	342819	1.514E+11
Yttrium	39	88.90590	1.2626	0.79202	339343	1.454E+11
Zirconium	40	91.22000	1.2634	0.79151	339451	1.456E+11
Niobium	41	92.90640	1.2491	0.80058	337524	1.423E+11
Molybdenum	42	95.94000	1.2672	0.78914	339961	1.465E+11
Technetium	43	98.96200	1.2843	0.77863	342247	1.504E+11
Ruthenium	44	101.07000	1.2799	0.78131	341660	1.494E+11
Rhodium	45	102.90550	1.2697	0.78759	340296	1.470E+11
Palladium	46	106.40000	1.2958	0.77172	343776	1.531E+11
Silver	47	107.86800	1.2779	0.78253	341393	1.489E+11
Cadmium	48	112.41000	1.3244	0.75506	347549	1.600E+11
Indium	49	114.82000	1.3258	0.75426	347732	1.603E+11
Tin	50	118.69000	1.3561	0.73741	351683	1.677E+11
Antimony	51	121.75000	1.3695	0.73019	353417	1.711E+11
Tellurium	52	127.60000	1.4356	0.69657	361845	1.880E+11
Iodine	53	126.90450	1.3766	0.72643	354332	1.728E+11
Xenon	54	131.30000	1.4134	0.70751	359037	1.822E+11
Caesium	55	132.90540	1.3985	0.71505	357139	1.784E+11
Barium	56	137.33000	1.4341	0.69730	361656	1.876E+11
Lanthanum	57	138.90550	1.4188	0.70482	359722	1.836E+11
Cerium	58	140.12000	1.3979	0.71536	357062	1.782E+11
Praseodymium	59	140.90770	1.3705	0.72966	353546	1.713E+11
Neodymium	60	144.24000	1.381	0.72411	354898	1.739E+11
Promethium	61	145.00000	1.3593	0.73567	352098	1.685E+11
Samarium	62	150.40000	1.4077	0.71038	358312	1.807E+11
Europium	63	151.96000	1.3942	0.71726	356590	1.773E+11
Gadolinium	64	157.25000	1.4387	0.69507	362236	1.888E+11
Terbium	65	158.92540	1.4268	0.70087	360735	1.857E+11
Dysprosium	66	162.50000	1.4438	0.69262	362877	1.901E+11
Holmium	67	164.93040	1.4433	0.69286	362818	1.900E+11
Erbium	68	167.26000	1.4414	0.69377	362575	1.895E+11
Thulium	69	168.93420	1.4301	0.69925	361151	1.865E+11
Ytterbium	70	173.04000	1.4536	0.68795	364107	1.927E+11
Lutetium	71	174.96700	1.446	0.69156	363154	1.907E+11
Hafnium	72	178.49000	1.4606	0.68465	364982	1.951E+11
Tantalum	73	180.94780	1.4603	0.68479	364945	1.599E+11
Tungsten	74	183.85000	1.466	0.68213	365656	1.960E+11
Rhenium	75	186.20700	1.4643	0.68292	365444	1.956E+11
Osmium	76	190.20000	1.484	0.67385	367894	2.009E+11
Iridium	77	192.22000	1.4778	0.67668	367125	1.992E+11
Platinum	78	195.09000	1.4826	0.67449	367721	2.005E+11
Gold	79	196.96650	1.4747	0.67810	366740	1.983E+11
Mercury	80	200.59000	1.4887	0.67173	368476	2.021E+11
Thallium	81	204.37000	1.5043	0.66476	370402	2.064E+11
Lead	82	207.80000	1.5153	0.65994	371754	2.094E+11
Bismuth	83	208.98040	1.4991	0.66707	369761	2.050E+11

Polonium	84	209	1.4695	0.68050	366093	1.969E+11
Astatine	85	210	1.4522	0.68861	363931	1.923E+11
Radon	86	222	1.5622	0.64012	377463	2.226E+11
Francium	87	223	1.5417	0.64863	374978	2.168E+11
Radium	88	226.0254	1.5494	0.64541	375914	2.189E+11
Actinium	89	227	1.5316	0.65291	373748	2.139E+11
Thorium	90	232.0381	1.559	0.64144	377076	2.217E+11
Protactinium	91	231.0359	1.52	0.65789	372330	2.107E+11
Uranium	92	238.0290	1.568	0.63776	**378163**	**2.242E+11**
Neptunium	93	237.0482	1.53	0.65359	373553	2.135E+11
Plutonium	94	244	1.5765	0.63432	379187	2.267E+11
Americium	95	243	1.5389	0.64981	374638	2.160E+11
Curium	96	247	1.5538	0.64358	376447	2.202E+11
Berkelium	97	247	1.5275	0.65466	373247	2.128E+11
Californium	98	251	1.5422	0.64842	375039	2.169E+11
Einsteinium	99	254	1.5466	0.64658	375574	2.182E+11
Fermium	100	257	1.5509	0.64479	376095	2.194E+11
Mendelevium	101	258	1.5355	0.65125	374224	2.150E+11
Nobelium	102	259	1.5293	0.65389	373467	2.133E+11
Lawrencium	103	256	1.4670	0.68166	365781	1.963E+11
Rutherforium	104	261	1.4909	0.67074	368749	2.027E+11
Dubnium	105	262	1.4767	0.67719	366988	1.989E+11
Seaborgium	106	263	1.4627	0.68367	365245	1.951E+11
Bohrium	107	262	1.4304	0.69911	361189	1.866E+11
Hassium	108	265	1.4354	0.69667	361820	1.879E+11
Meitnerium	109	266	1.4222	0.70314	360153	1.845E+11
Darmstadtium	110	271	1.4453	0.69190	363066	1.905E+11
Roentgenium	111	272	1.4322	0.69823	361417	1.871E+11

Table
1.1

* the speed value above the Earth surface is increased.

Table 1.1

S.

1.2

1.2

The quantum speed and the process/anti-process relation

1F1   The electron process intensity drive of the *direct process* is: $e = \dfrac{dmc^2}{dt_i \varepsilon}\left(1 - \sqrt{1 - \dfrac{(c-i)^2}{c^2}}\right)$;

$\dfrac{dmc^2}{dt_i \varepsilon}$ For simplicity only this part of the equation will be analysed, as the other components are standard ones, in fact equal and they are not influencing the outcome.

1F2   The source (taken here as drive) of the *anti-process* is: $\dfrac{dmc^2}{dt_i \varepsilon_-} = \dfrac{dmc^2}{dt_i dt_{p-}} dt_{n-}$

Comparing the direct and the anti-processes:

➤ The speed of quantum communication should be one and the same, since this is one and the same for the given elementary process.

➤ The unified process intensities for the direct and the anti-processes are different.

➤ This is not contradicting to the just proven constancy of the values, since the intensity of the anti-electron process in 1D4 was deliberately modified to ε_{-xm} .

While $c = c_-$ the intensities are $\dfrac{dt_n}{dt_p} \neq \dfrac{dt_{p-}}{dt_{n-}}$; and $\varepsilon \neq \varepsilon_-$ 1F3

The intensity of the drive of the collapse of the anti-proton process is different to the intensity of the drive of the neutron process:	$\varepsilon = \dfrac{1}{\varepsilon_-}$; as	$dt_n = dt_{-n}$ and $dt_{-p} = dt_p$	and $\varepsilon_- = \dfrac{dt_{-p}}{dt_{-n}}$	1F4

As the speed values of the quantum communication of the direct and the anti-processes should be equal, the denominator of the *IQ* of the drive and the anti-drive (the quantum speed) are the same, but the values of the nominator (the intensities) must be different.

➤ In the case of proton process dominant elementary processes, the generation of the *blue shift* impact of the electron process is more than the use, therefore, there is a surplus generating on the direct side with intensity value less than the intensity of the anti-electron process.

➤ In the case of neutron process dominant elementary processes, the electron process surplus is generating on the anti-process side, with intensity of the *blue shift* impact less than the intensity of the electron process on the direct side.

There is a surplus in the process: $\Delta n = n_{generating} - n_{use}$ in both cases.   1F5

➤ in the first case, the surplus is used for initiating elementary communication;
➤ in the second, for feeding *gravitation*.

The surplus without use generates electron process *blue shift* conflict. In this case, the quantum membrane becomes of increased intensity with increased speed of quantum communication.

The increase of the quantum speed for any reason also generates conflict.

Nature is taking care of the use in both versions.

The difference in the intensities of the electron and the anti-electron processes and, at the same time, the equality of the quantum speed values mean that the unified process intensity values in the direct and the anti-processes are not equal: $c^2 \cdot \varepsilon \neq c^2 \cdot \varepsilon_-$ 1F6

The equation, however, can be written in different way: $c^2 \cdot \varepsilon = \dfrac{c^2}{\varepsilon_-}$ 1F7

1F6 and 1F7 describe the relations of the direct and the anti-processes with and without taking into account the surplus.

➤ the speed value relates to the normal use, meaning the surplus is utilised in its natural way: either for *gravitation* (in neutron dominant elementary processes) or for initiating elementary communication (in proton dominant elementary processes); this note is important as the not used surplus increases the speed of quantum communication.

➤ the intensity values of the electron processes represent the generating intensities within the direct and the anti-processes; there is no difference, whether they are in surplus, or whether the surplus is being used or not.

If the surplus has not been utilised 1F7 loses its validity: $c^2 \cdot \varepsilon \neq \dfrac{c^2}{\varepsilon_-}$ There is a conflict with no difference which side is the one without the utilisation of the surplus.   1F8

The cyclical character of the elementary processes of the evolution and the *entropy* mean that the equation in 1F7 for the elementary processes are never exactly equal.

This is the reason the equality in 1F7 shall be considered as valid but in fact taken as equal to $c^2 \cdot \varepsilon \cong \dfrac{c^2}{\varepsilon_{-m}}$ 1F9

As a result of the rule of the elementary evolution from *plasma* to the *Hydrogen* process, the proton process after the *inflexion* has increasing intensity from each of the steps.

Plasma is electron process *blue shift* impact of infinite high intensity and anti-electron process *blue shift* surplus in infinite volume. The surplus of the anti-electron process *blue shift* impact is getting less and less by the progress,

1G1   meaning: $\varepsilon_{-m} < \varepsilon_{-}$ It means: $\dfrac{1}{\varepsilon_{-m}} > \dfrac{1}{\varepsilon_{-}}$ And as $\varepsilon_{-} = \dfrac{dt_{p-}}{dt_{n-}}$

1G2

$\varepsilon_{-m} < \varepsilon_{-}$ means the intensity growth of the anti-proton collapse.

The anti-electron process surplus and the quantum membrane on the anti-process side make possible the growth.

1G3   If it is taken that $\varepsilon_{next} = \dfrac{\varepsilon_{p-next}}{\varepsilon_{n-next}} = \dfrac{dt_{n-next}}{dt_{p-next}}$; and $dt_{p-next} < dt_{p-previous} \Rightarrow \varepsilon_{next} > \varepsilon_{previous}$

the intensity of the electron process of the following cycle will be of higher value.

The intensity of the proton expansion of the next (following) elementary cycle is equal to the intensity of the anti-proton collapse of the previous cycle, because the intensities of the two sides of the *inflexion* are equal.

This is not really an intensity growth, rather, an increase relative to the neutron process; the increase of the proton process dominance. The intensity of the neutron process is getting less and less within each cycle. This is understandable since the original infinite high intensity of the *plasma* is working out as it evolves. The neutron process is driven and "neutral". The speed of the quantum drive in each elementary cycle is getting less and less.

The quantum speed at *plasma* is infinite high, the quantum speed of the *Hydrogen* process is infinite low and all elementary processes between have quantum speed value with decreasing gradient.

1G4   1F7 can be developed as: $c^2 \cdot \varepsilon \cong \dfrac{c^2}{\varepsilon_{-m}} \cong c^2_{next} \cdot \varepsilon_{next} \cong \dfrac{c^2_{next}}{\varepsilon_{-nextm}} \cong \ldots$

The unified process and anti-process intensity relations have the global view of:

anti-proton / proton inflexion

1G5

$$c^2_{previous} \cdot \varepsilon_{previous} \cong \dfrac{c^2_{previous}}{\varepsilon_{-previous/m}} = c^2_{next} \cdot \varepsilon_{next} \cong \dfrac{c^2_{next}}{\varepsilon_{-next/m}} = \ldots = c^2_n \cdot \varepsilon_n \cong \dfrac{c^2_n}{\varepsilon_{-n/m}} = c^2_{n+1} \cdot \varepsilon_{n+1}$$

elementary cycle

1G6 $\dfrac{1}{\varepsilon_{-previous/m}} = \varepsilon_{next}$ and $\varepsilon_{next} > \varepsilon_{previous}$ at the same time: $c_{next} < c_{previous}$

the two different gradients are compensating each other

The quantum speed is decreasing; the intensity coefficient of the direct process is increasing: the elementary evolution is on its way to proton process dominancy.

As the intensity of the anti-neutron process through the *inflexion* strongly follows the intensity of the neutron process, the energy intensity source of the anti-electron process, the drive of the anti-proton collapse is also decreasing.

The *inverse* of the unified process and anti-process intensities in 1G5 does not work:

1G7 $\dfrac{c^2_{previous}}{\varepsilon_{previous}} \cong c^2_{previous} \cdot \varepsilon_{-previous/m} \neq \dfrac{c^2_{next}}{\varepsilon_{next}} \cong c^2_{next} \cdot \varepsilon_{-next/m} \neq \ldots \neq \dfrac{c^2_n}{\varepsilon_n} \cong c^2_n \cdot \varepsilon_{-n/m} \neq \dfrac{c^2_{n+1}}{\varepsilon_{n+1}} \cong \ldots$

By the growth of the intensity coefficient (ε) and by the decrease of the quantum speed (c) the quotients of the two are not equal within the cycles. This quotient gives the IQ value,

with decreasing gradient as presented in Table 1.1.
- Elementary processes with *neutron process dominance* are the source of *gravitation*;
- Elements with *proton process dominance* are the source of elementary communication.

Neutron process dominancy, in fact, is the consequence of the increased intensity of the direct drive of the collapse. The surplus of the anti-electron process *blue shift* impact – as the elementary control of the direct process - is correcting the balance, loading *gravitation*, driving the elementary evolution.

Proton process dominant elementary processes are the drives of elementary communication: looking for elementary communication for using the available electron process *blue shift* surplus. The anti-process in fact "accumulates" the energy source of the direct process.

Processes and anti-processes are not in cycle.
The process and the controlling anti-process run in parallel.
While the direct and anti-processes run in parallel, the process is about two *inflexions*:
➢ at the end of the collapse leading to the expansion; and
➢ at the end of the expanded state with the start of the collapse.
The *entropy* of the processes and the generating quantum impulse – with reference to Section 2 establish, build up and feed the quantum system of reference.

The step by step modification of the intensities of the process and anti-process chain is producing an infinite number of elementary processes with fully balanced and partially balanced (isotope) $c^2 \cdot \varepsilon = const$ relations.

1.3.

The quantum impact of *gravitation*

S.

1.3

Under the quantum impact of *gravitation*, elementary processes are functioning by c, the quantum speed of *gravitation* (gaseous state or state after melting).
The elementary process structure becomes rearranged:

the original in minerals	*the element from mineral state*	*the relations*		Ref.
$\dfrac{dmc_x^2}{dt_i \varepsilon_x}\left(1-\sqrt{1-\dfrac{(c_x-i_x)^2}{c_x^2}}\right)$	$n\dfrac{dmc^2}{dt_i \varepsilon_x}\left(1-\sqrt{1-\dfrac{(c-i)^2}{c^2}}\right)$	$\begin{aligned}dmc_x^2 =\\= n \cdot dmc^2\end{aligned}$	$n=\dfrac{c_x^2}{c^2}$	1D2 1H1

For *Nitrogen* and *Oxygen* n is equal to 0.9925 and 0.9980; for *Uranium*, *Lead* and *Mercury*, the heaviest elements to 1.5900; 1.5355 and 1.5086 respectively.

➢ From 1H1 it follows that elementary processes with higher quantum speed than c, once transformed to speed value c will have more elementary process cycles. More elementary cycles generate more *blue shift* impact and increase the intensity:

$$n\frac{dmc^2}{dt_i \varepsilon_x}\left(1-\sqrt{1-\frac{(c-i)^2}{c^2}}\right)=\frac{dmc^2}{dt_i \dfrac{\varepsilon_x}{n}}\left(1-\sqrt{1-\frac{(c-i)^2}{c^2}}\right);$$

It is equivalent to an elementary process with less quantum speed and more intensity.

1H2

This is the reason for the softened status.
➢ In the case of $n<1$, the gaseous and $n \ll 1$, the volatile gaseous states mean: the *blue shift* surplus and, in this way, the increased *blue shift* conflict are acting. The quantum speed is higher than it is necessary for the *inflexion*.
There is no way fractions of elementary cycles would work.

The *Quantum Membrane* becomes overloaded, because the quantum speed of the *inflexion* should be of less value. The only way the standard *inflexion* of the elementary structure could correspond to this increased speed value is if the element is speeded up:

1H3
1H4

Once the element is speeded up to speed v

$$n \frac{dmc^2}{dt_i \varepsilon_x \sqrt{1 - \frac{v^2}{c^2}}} \left(1 - \sqrt{1 - \frac{(c-i)^2}{c^2}}\right); \qquad \frac{n}{\sqrt{1 - \frac{v^2}{c^2}}} = 1; \qquad v = c\sqrt{1 - n^2}$$

The speed increase in the case of *Nitrogen* takes up to 36,590 km/sec. This further increases the conflict.

S.
1.4

1.4
Hydrogen process and *gravitation*

1I1

Hydrogen process as result of elementary evolution means an elementary process with intensity coefficient of infinite high value:

$$\lim \varepsilon_{eH} = \lim \frac{\varepsilon_p}{\varepsilon_n} \sqrt{1 - \frac{(c-i)^2}{c^2}} = \infty$$

1I2
1I3

As the intensity coefficient is $\lim \varepsilon_{eH} = \infty$	the *Intensity Quotient (IQ)* – the relation of the two is:	$\lim \dfrac{c_H^2}{\varepsilon_{eH}} = 0$

The "natural," generated by *gravitation, Hydrogen* process becomes part of the *quantum system* of the *Earth* with speed of quantum communication:

Ref.

1H1

$$c = 299{,}792 \, \text{km/sec.}$$

1I4

With reference to 1H1, n is of infinite low value, as $\lim n = \lim \dfrac{c_H^2}{c^2} = 0$

1I1

The electron process of the natural *Hydrogen* is: $\quad e_H = \dfrac{dmc_H^2}{dt_i \varepsilon_{eH}} \left(1 - \sqrt{1 - \dfrac{(c_H - i_H)^2}{c_H^2}}\right)$

1I2

The electron process *blue shift* drive of this natural *Hydrogen*, on the surface of the *Earth*, with increased quantum speed and *blue shift* conflict, as of 1H4 is:

$$e_H = n \frac{dmc^2}{dt_i \varepsilon_{eH}} \left(1 - \sqrt{1 - \frac{(c-i)^2}{c^2}}\right)$$

The *blue shift* conflict in 1I2 is not significant, because:

➤ ε_{eH} the intensity coefficient still remains of infinite high value; dt_i is the time system on the *Earth* surface;

➤ *gravitation* is the impact of: $g\Delta t = c - i$; $c = 299{,}792$ km/s in 1H2 corresponds to the speed value of the quantum communication on the surface of the *Earth*, $i = \lim a\Delta t = c$ (*Oxygen, Nitrogen* and *Helium* processes behave the same way above the *Earth* surface.) n is of infinite low value – therefore, finding *Hydrogen* in *Earth* atmosphere is practically impossible.

1I3

If *Hydrogen* process is accelerated on the surface of the *Earth*, the internal conflict is increasing, as the resulting energy intensity of the electron process is increasing:

$$e_H = \frac{dmc^2}{dt_i \varepsilon_{eH} \sqrt{1 - \frac{v^2}{c^2}}} \left(1 - \sqrt{1 - \frac{(c-i)^2}{c^2}}\right)$$

Sufficiently high speed value of v of the acceleration,
has its intensity increasing effect – at the intensity – indeed: $\lim\sqrt{1-\dfrac{v^2}{c^2}}=0$ 1I4

The elementary process of the *Hydrogen* in acceleration within an accelerating channel is subject to *blue shift* impact of *gravitation* as well. The benefit of the acceleration is that with the increasing speed of the acceleration, approaching c, the *blue shift* conflict within the accelerating channel is increasing dramatically.
The conflict can be summarised in the following way:

$$e_{H\Sigma}=\frac{dmc^2}{dt_i\varepsilon_{eH}\sqrt{1-\dfrac{v^2}{c^2}}}\left(1-\sqrt{1-\dfrac{(c-i)^2}{c^2}}\right)+\frac{dmc^2}{dt_i\varepsilon_g}\left(1-\sqrt{1-\dfrac{(c-i)^2}{c^2}}\right);\quad \text{here } \varepsilon_g=1 \qquad 1I5$$

$$e_{H\Sigma}_{\lim v=i}=\frac{dm}{dt_i}\left(\frac{c^2}{\varepsilon_{eH}\sqrt{1-\dfrac{v^2}{c^2}}}+\frac{c^2}{\varepsilon_g}\left(1-\sqrt{1-\dfrac{(c-i)^2}{c^2}}\right)\right) \qquad 1I6$$

$\varepsilon_g=1$ in 1I5 is coming from the fact that all elementary processes with $\varepsilon_x>1$ are leaving the *Earth* surface, while all others with $\varepsilon_x<1$ are part of it.
The acting *blue shift* conflict causes heat generation.
Heat generation is a source of energy!

Does it make any difference what the elementary subject of the acceleration is?
The difference is significant!

while for the $\lim\dfrac{c^2}{\varepsilon_{eH}}=0$ for any other elementary relations, where $\dfrac{c^2}{\varepsilon_x}\gg 1$ 1I7
Hydrogen $c_x\ge 299{,}792$ km/sec and $\varepsilon_x\le 1$

The acceleration for all other elements would need infinite high energy!
Therefore, *Hydrogen* is the only reasonable and practical option.

1.5   S.
The conflict with the intensity of the *gravitation* on the *Earth* surface   1.5

The intensity and, thus, the effect of *gravitation* is of infinite low value *for us* on the surface of the *Earth*.
The reason is the quasi constant speed of our motion on the *Earth* surface (taken by the sphere symmetrical expanding acceleration of the *Earth*) by $i=\lim a\Delta t=c$.
Elementary processes in gaseous state (like *Oxygen, Nitrogen, dioxides* or other products of industrial use) have also been "taken" by *Earth* surface by their electron process *blue shift* conflict with *gravitation*.

The energy intensity of $e_g=\dfrac{dmc^2}{dt_i\varepsilon_g}\left(1-\sqrt{1-\dfrac{(c-i)^2}{c^2}}\right)$ where dt_i and ε_g are the time   1J1
the acting *gravitation* is: system and the intensity
                                                                                               coefficient of *gravitation*   1J2

The energy intensity of $e_{gases}=\dfrac{dmc^2}{dt_i\varepsilon_{gases}}\left(1-\sqrt{1-\dfrac{(c-i)^2}{c^2}}\right)$ The time system and the   Ref.
the gaseous state is: quantum speed are the   Table
                                                                                               same   1.1

With reference to Table 1.1: $\varepsilon_{gases} > \varepsilon_g$ is the reason for gaseous state, the reason for the conflict.

1J3

The absolute energy of the acting *gravitation* (for a certain period) would be

$$E_g = \frac{dmc^2}{dt_i \varepsilon_i \varepsilon_g}\left(1 - \sqrt{1 - \frac{(c-i)^2}{c^2}}\right)$$

The intensity might be taken as reciprocal to the timeframe $\varepsilon_i = 1/dt_i$

c, the speed of quantum communication on the *Earth* surface, has been established by the acceleration of the *Earth*. The loading *blue shift* impact of the quantum system of the *Earth* is *gravitation*.

The indexes of the time and the intensity coefficients in 1J1 are taken as different.

The reason is that ε_g, value of $\varepsilon_g = 1$, with reference to 1I5, is for the characterisation of the intensity of the "electron function" (the proton/neutron relation) of *gravitation*:

1J4

The proton and the neutron functions of *Earth gravitation* are not known, but the value of the intensity coefficient of *gravitation* is taken with reference to 1I6, as:

$$\varepsilon_g = \frac{\varepsilon_{pg}}{\varepsilon_{ng}}\sqrt{1 - \frac{(c-i)^2}{c^2}} = 1$$

We live on the *Earth* surface "at relative rest" within dt_E time system while, in fact, having been taken by the sphere symmetrical expanding acceleration of the *Earth*, *gravitation*.

While our social time count is specific and relates to the rotation of the *Earth*, the time system of reference is the one of *Earth* expanding acceleration: $dt_E = dt_i$

Our rest status corresponds to the sphere symmetrical expanding acceleration of the *Earth* surface, the motion with $i = \lim g\Delta t = c$. Similarly to solid subjects, elementary processes, separated from *Earth* structure (not being component to *gravitation* any more) – are taken as well. Water sources and liquids are the ones between: conflicting and taken. In fact, solid structures fully represent and "participate" in *gravitation*; liquid and gaseous partially.

1J5

The energy intensity of *gravitation* (the *blue shift* impact) in 1J1 is acting within the time system of the *Earth* surface (our life)

$$e_E = \frac{dmc^2}{dt_E \varepsilon_g}\left(1 - \sqrt{1 - \frac{(c-i)^2}{c^2}}\right)$$

Ref. 1A4 1J6

With reference to 1A4, *Earth* surface is in quantum communication with *plasma*: even the quantum speed at the *plasma* is $\lim c_{pl} = \infty$ and the intensity coefficient of the electron process is of $\lim \varepsilon_{pl} = 0$

$$Z_{plasma} = \frac{1}{\varepsilon_{pl}} = \frac{dt_p}{dt_n} = \infty$$

= *event concentration* is of infinite high value

The quantum message of the *plasma* state is twofold:

1K1

1: the intensity of the electron process and the *blue shift* conflict of the *plasma* is of infinite value:

$$e_{plasma} = \frac{dmc_{pl}^2}{dt_i \varepsilon_{pl}}\left(1 - \sqrt{1 - \frac{(c_{pl} - i_{pl})^2}{c_{pl}^2}}\right)$$

1K2

2: even in the case of infinite value of event concentration and infinite value of proton process intensity the form of $i_x = \lim a_x \Delta t = c_x$ guaranties and establishes communication

$$dt_i = \frac{dt_o}{\sqrt{1 - \frac{i_x^2}{c_x^2}}}$$

With the quantum message of the *Hydrogen* process at the other end:
energy/mass/energy... change is constant and forever.

For having *blue shift* conflict with *Earth gravitation* we should either
> slow down our time system, increasing by that, the intensity of the *Quantum Membrane* above the *Earth* surface, which is impossible since the *Quantum Membrane* has been established by *gravitation* itself; or
> speeding up elementary processes, slowing down (lengthening by that) the time-flow, for having *blue shift* impact and conflict with *gravitation*. All elementary processes on the *Earth* surface having electron process intensity coefficient $\varepsilon_x > 1$ (solid and liquid status) do not have conflict, since their intensities are prevailing.

For being in *blue shift* conflict with *gravitation* or with other elementary processes, elements have to be in conflict first within their own elementary structure.
The first condition of any conflict is *blue shift* surplus.

The speeded-up electron process means increased *blue shift* surplus:

$$e_{x-speeded-up} = \frac{dmc^2}{dt_i \varepsilon_x \sqrt{1 - \frac{v^2}{c^2}}} \left(1 - \sqrt{1 - \frac{(c_x - i_x)^2}{c_x^2}} \right) = n \frac{dmc^2}{dt_i \varepsilon_x} \left(1 - \sqrt{1 - \frac{(c_x - i_x)^2}{c_x^2}} \right); \quad n = \frac{1}{\sqrt{1 - \frac{v^2}{c^2}}} \qquad 1K3$$

The speeded-up *blue shift* impact is increasing the surplus – basis for conflict – since the intensity quotient *IQ* of the speeded-up element remains the same.

Gravitation is acting through any elementary structures having been or not in conflict. Structures themselves can also be subject to conflict, with symptoms of increasing temperature. (Liquid structures could, therefore, be transforming into gaseous state.)

1.6

From *plasma* state to the *Hydrogen* process

The speed of quantum communication is changing from *Earth plasma* towards the surface.
Plasma is *blue shift* conflict of infinite high intensity.
The infinite quantum speed and the infinite *IQ* value of the *plasma* results in infinite anti-electron process *blue shift* conflict.

Plasma and all elementary processes of the core give off their anti-electron process *blue shift* surplus, feeding by that *gravitation*. dt_i the time systems of all electron process *blue shift* impacts released towards the *Earth* surface are quasi equal:

$$dt_i = \frac{dt_o}{\sqrt{1 - \frac{i^2}{c^2}}} = \frac{dt_o}{\sqrt{1 - \frac{\lim^2 c}{c^2}}};$$

The quasi equality is valid for all calculated quantum speed values, including $\lim c_H = 0$ for the *Hydrogen*. This equation is the proof of elementary communication.

1L1

Losing on the intensity of the *blue shift* impact towards the *Earth* surface, the original anti-electron process *blue shift* conflict is decreasing, the core is formulating.
This also means the decrease of c_x the speed value of quantum communication and the decrease of the intensity of the electron process (the increase of ε_x, the intensity coefficient).

1L2 The *intensity coefficient* of the electron process is the reciprocal value of the *event concentration*:

$$\varepsilon_x = \frac{1}{Z_x}$$

Z_x is of infinite high value in *plasma*: $\lim Z_{pl} = \infty$ and this way $\lim \varepsilon_{pl} = 0$

1L3
$$\frac{\Delta m \left(\frac{c_{pl}}{x}\right)^2}{\Delta t (\varepsilon_{pl} \cdot y)}\left(1 - \sqrt{1 - \frac{(c-i)^2}{c^2}}\right) = \frac{\Delta m c_x^2}{\Delta t \varepsilon_x}\left(1 - \sqrt{1 - \frac{(c-i)^2}{c^2}}\right)$$

The *blue shift* conflict of the *plasma* needs consolidation. The energy surplus shall be given off.

We have to stop here to explain the case.

Plasma is *inflexion*, collapse of infinite high intensity, result of the infinite high intensity of the *IQ* value of the electron process impact – start of *expansion*.

ε_x the intensity coefficient is increasing, just leaving the $\lim \varepsilon_{pl} = 0$ value and

c_x the speed of quantum communication is decreasing, but the *plasma* conflict is still on as the *IQ* is decreasing step by step by the *expansion*.

> (The lesson is that *plasma*, fire, melting and other similar statuses are the result of the infinite high value of the *IQ* of the electron process. This means either the quantum speed is of infinite high value itself, or the intensity of the *blue shift* impact generates conflict, which results in increased – destroying the elementary structure – speed of quantum communication.)

x and y characterise the modification of the quantum speed and the intensity of the electron process. With the loss by *gravitation* of the anti-electron process *blue shift* conflict, the speed values and the intensities of elementary processes are changing: nature creates elements with certain and specific characteristics in line with the cooling effect of the *plasma* (by *gravitation*) as in 1K3.

x and y are results of the cooling by *gravitation* (at *plasma* state $x = 1$ and $y = 1$):

1L4
$$dg = \frac{dm}{dt_i}\frac{c_{pl}^2}{1 \cdot \varepsilon_{pl} \cdot 1}\left(1 - \sqrt{1 - \frac{(c_{pl} - i_{pl})^2}{c_{Pl}^2}}\right) - \frac{dm}{dt_i}\frac{c_{pl}^2}{x^2}\frac{1}{\varepsilon_{pl} y}\left(1 - \sqrt{1 - \frac{(c_{pl} - i_{pl})^2}{c_{pl}^2}}\right)$$

As proof of the cooling impact of the surplus of the anti-electron process of elementary processes and the transformation of the intensity and quantum speed of the *plasma* state to the *Hydrogen* process, the heat flows constantly from its sources within the *Earth* to the surface. Total heat loss from the *Earth* is estimated at 44.2 TW.

The geothermal gradients are:

   65 mW/m² heat flow over continental crust and 101 mW/m² over oceanic crust.
   [Source: Pollack, Henry N., et.al., *Heat flow from the Earth's interior.* August 1993.]

This is equal to the intensity of the *blue shift* impact of *gravitation*, leaving behind always quantum speed c_x of less value with certain increasing intensity coefficient (ε_x) of the process.

1L5
$$dg = \frac{dm}{dt_i}\frac{c_{xn}^2}{\varepsilon_{xn}}\left(1 - \sqrt{1 - \frac{(c-i)^2}{c^2}}\right) - \frac{dm}{dt_i}\frac{c_{x(n+1)}^2}{\varepsilon_{x(n+1)}}\left(1 - \sqrt{1 - \frac{(c-i)^2}{c^2}}\right)$$

With the *gravitation* going on, *plasma* status, as the origin and the centre of the process is shrinking in its size; the speed value of the *Earth* surface, as the external boundary of the core is decreasing.

The last elementary status of the cooling process is with proportions $x = \infty$ and $y = \infty$.

These indicators for the elementary process mean: $\lim c_x = 0$ and $\lim \varepsilon_x = \infty$.

The cooling by *gravitation* is in effect and the balance is managed:

$$mc_{pl}^2 - \Delta = 0; \quad \Delta = mc_{pl}^2$$

1M1

Meaning: the mass/energy impact of the *plasma* state with $\lim c_x = \infty$ is transforming into a different state with $\lim c_x = 0$ (*Hydrogen* process). There is an infinite number of balance statuses between these two ends of the transformation, what we found as elements, elementary processes.

If we describe the intensity of the *plasma* process (giving off energy) as $\dfrac{dmc_{pl}^2}{dt_i \varepsilon_{pl}}\left(1 - \sqrt{1 - \dfrac{(c_{pl} - i_{pl})^2}{c_{pl}^2}}\right)$ with $\lim c_{pl} = \infty$ and $\lim \varepsilon_{pl} = 0$

1M2

The energy intensity of *gravitation* from *plasma* to *Earth'* surface gives:

$$\frac{dmc_{pl}^2}{dt_i \varepsilon_{pl}}\left(1 - \sqrt{1 - \frac{(c_{pl} - i_{pl})^2}{c_{pl}^2}}\right) > ... > \frac{dmc_x^2}{dt_i \varepsilon_x}\left(1 - \sqrt{1 - \frac{(c_x - i_x)^2}{c_x^2}}\right) > \frac{dmc^2}{dt_i \varepsilon_{xE}}\left(1 - \sqrt{1 - \frac{(c - i)^2}{c^2}}\right)$$

1M3

with $\lim \varepsilon_{pl} = 0$; $\varepsilon_E = 1$ and for all elementary processes: $\varepsilon_x^n < \varepsilon_{xE} < \varepsilon_x^p$ with $\lim \varepsilon_H = \infty$.

1M3 can also be interpreted as: $dt_x = dt_i \varepsilon_{xE}$ and $dt_H = dt_i \varepsilon_H$

1M4

(There is no conflict in 1M4, the intensity of the electron process is a coefficient without dimension.)

The result for each elementary process this way may give the meaning of *time*: $\dfrac{dmc_x^2}{dt_i \varepsilon_x} = \dfrac{dmc_x^2}{dt_x}$

1M5

$dt_x = dt_i \cdot \varepsilon_x = \dfrac{dt_i}{Z_x}$ The equal time system of all electron process *blue shift* impacts and the value of event concentration give the time system of the element.

1M6

Time systems are about the intensities of elementary processes.

Time system dt_x means being part of this process.

1.7
Space-time has been established by the speed of quantum communication

S.
1.7

Looking for the distance measurement as the characteristic of the space of the quantum systems, the obvious and simple formula is: $ds = vdt$

1N1

For the measurement of the distance of an electron process *blue shift* impact of certain quantum speed and intensity, 1N1 will have the format of $ds_x = c_x dt_i \varepsilon_x$

1N2

- c_x is understandable in 1N2, since this is indeed the speed of the quantum signal;

- dt_i is the common time system of electron process *blue shift* impacts; therefore, using it in the formula seems correct – the distance is speed effect in the acting time system;

- the *blue shift* impacts of equal timeframe in elementary quantum systems, however, have their definite function;

- this is to drive the neutron processes;

- therefore, the category of distance shall be connected to the function and it relates to:

1N3 $\dfrac{dmc_x^2}{dt_i \varepsilon_x}$ the *dmc*^{2} change occurs in fact for $dt_i \varepsilon_x$ time period.

Ref.   -   Therefore, the functional timeframe of the electron process *blue shift* impact in
1N2   quantum systems, with reference to 1N2, is $dt_i \varepsilon_x$ indeed.

Space-time quantum systems are loaded by electron process *blue shift* impacts.
Earth space-time is loaded by the *blue shift* impact of *gravitation*, speed of $c_E = 299,792$ km/sec, intensity coefficient of $\varepsilon_E = 1$.
The value of the quantum speed above the *Earth* surface is established by the *quantum impact* of *gravitation*.
The *approaching impact* of *gravitation* is the sphere symmetrical expanding acceleration of the *Earth* surface by $i_E = \lim g \Delta t = c_E$.

For developing the case the formula in 1N2 above shall be multiplied and divided by c_x.

1N4   Rearranging the formula it gives: $ds_x = \dfrac{c_x^2 \cdot \varepsilon_x}{c_x} dt_i$; and $\dfrac{ds_x}{dt_i} = \dfrac{c_x^2 \cdot \varepsilon_x}{c_x}$

1N4 above gives the proof of the considerations about the distance of space-times:
-   the unified process intensity is constant to all elementary processes, as equal and constant the electron process time system as well.
-   the distance as space parameter depends on the speed of the quantum impacts.

Once with reference to the earlier sections:

1N5 $c_x^2 \cdot \varepsilon_x = const$; and $dt_i = const$ $ds_x = A \cdot f \left(\dfrac{1}{c_x} \right)$
              the change of the space coordinate is function of the
                           quantum speed value

1N4 and 1N5 mean: space is characterised by the *speed value of quantum communication*. The higher the quantum speed is, the less is the change of the space coordinate. This is not equivalent to being small. On the contrary, this is about events in space and time of increased intensity, the concentration of space and time.

We can formulate 1N5 in different way:
-   while the quantum impact with $\lim c_x = \infty$ effects $\lim n = \infty$ number of quantum impulses (quantum) – which means an unlimited large *space*;
-   quantum impact with $\lim c_x = 0$ effects only a limited, $\lim n = 0$ number of quantum – a limited *space*!

Space is about the intensity of the quantum system acting within. Therefore, the precise definition is *space-time* indeed.

2

The *quantum impulse (energy quantum)*

The proof of the $c^2 \cdot \varepsilon = const$ statement with reference to formula 1D3 in Section 1.1 is guaranteed by the cyclical character of the elementary evolution, the entropy generation of the elementary processes and the existence of the anti-processes.

The quantum speed value of the elementary process and the anti-process is one and the same, an equal value, but the intensities of the quantum drives are different. With reference to 1G1, 1G2 the intensity of the anti-electron process is a modified value. The reason is that the anti-proton collapse and the following proton expansion of the elementary processes should be equal. Otherwise, the elementary balance would be disrupted.

It means, with reference to the explanation in Section 1.1 and to the formula in 1F4, that

$$\varepsilon_n = \varepsilon_{n-} \text{ and } \varepsilon_{p-} = \varepsilon_p \text{ ; and also}$$

$$dt_n = dt_{n-} \text{ and } dt_{p-} = dt_p$$

ε_{m-} - is the modified intensity coefficient of the anti-electron process. The anti-electron processes are in surplus and the difference is feeding *gravitation*.

The *IQ* for the anti-electron process is: $IQ_- = \dfrac{c^2}{\varepsilon_{m-}}$; and $\varepsilon_{m-} = \dfrac{\varepsilon_{n-}}{\varepsilon_{p-}}\sqrt{1 - \dfrac{(c-i)^2}{c^2}} = \dfrac{dt_{p-}}{dt_{n-}}$;

The intensity of the electron process: $\varepsilon = \dfrac{\varepsilon_p}{\varepsilon_n}\sqrt{1 - \dfrac{(c-i)^2}{c^2}} = \dfrac{dt_n}{dt_p}$;

- the natural *entropy* impact and the *quantum impulse* generation,
- the relations in 2A1 and 2A2 above about $\varepsilon_n = \varepsilon_{n-}$ and $\varepsilon_p = \varepsilon_{p-}$,
- the equality of the quantum speed of the direct and the anti-processes mean:

$$IQ_- = \left|\frac{c_{previous}^2}{\varepsilon_{m-}^{previous}}\right| \cong c_{previous}^2 \cdot \varepsilon_e^{previous} = c_{following}^2 \cdot \varepsilon_e^{followig} \qquad \begin{array}{l}\text{'}\cong\text{' in the formula is explained by the}\\ \textit{entropy} \text{ effect of the elementary cycle.}\end{array}$$

and continuing: $c_{actual}^2 \cdot \varepsilon_{actual} \cong \left|\dfrac{c_{actual}^2}{\varepsilon_{m-}^{actual}}\right| = c_{following}^2 \cdot \varepsilon_e^{followig}$ ... and so on

With reference to 1D3 $c^2 \cdot \varepsilon = const$ indeed. This also means that the anti-electron process quantum drives of all anti-processes are equal!

The only concern could be about the 8 elementary processes with proton process dominance: as the electron process intensity surplus is generating on the direct side.

For these elementary processes, however: $\varepsilon_- = \varepsilon_{m-}$

For calculating the real intensity drive of the neutron process, the correction shall be made on the direct side: $\varepsilon_{m+} = \varepsilon - \varepsilon_s$; $-\varepsilon_s$ denotes here the intensity surplus for the elementary communication (as it was for the anti-processes ε_g for feeding *gravitation*).

Therefore, the equality for the 8 proton process dominant elementary processes shall be written as: $\left|\dfrac{c^2}{\varepsilon_-}\right| = c^2 \cdot \varepsilon_{m+}$

The general formula for all elementary processes can be written:

2A9
$$\frac{c_{x1}^2}{\varepsilon_{x1-}} = \frac{c_{x2}^2}{\varepsilon_{x2-}} = ... = c_{y1}^2 \cdot \varepsilon_{y1} = c_{y2}^2 \cdot \varepsilon_{y2} = ... = const$$

x-denotes proton process dominant elementary processes
y-denotes neutron process dominant elementary processes

The data in Table 1.1 has been calculated on the $c^2 \cdot \varepsilon = const$ basis. The difference is not significant.

The time system of dt_i, the quantum communication is one and the same for all electron processes. The intensities of the quantum impacts, however, vary.

The *IQ* of the electron process *blue shift* drive is the distinctive feature of each elementary process. The only way to modify it is to change the number of the acting impacts, create conflict (reference to Section 1.2) – a new space-time:

2B1
$$e_{imp} = n_x \frac{dmc_x^2}{dt_i \varepsilon_x}\left(1 - \sqrt{1 - \frac{(c_x - i_x)^2}{c_x^2}}\right)$$

ε_x is the intensity of the quantum impact;

n_x is the number of the impacts;

c_x is the quantum speed of the impact.

2B1 can be written also in the form of

2B2
$$e_{imp} = \frac{dmc_x^2}{dt_i \frac{\varepsilon_x}{n_x}}\left(1 - \sqrt{1 - \frac{(c_x - i_x)^2}{c_{x2}}}\right);$$

The *IQ* quantum drive of the conflict is: $IQ_{imp} = n\frac{c_x^2}{\varepsilon_x} = \frac{c_x^2}{\frac{\varepsilon_x}{n}}$

2B2 demonstrates that the intensity of the quantum impact depends on the relation of the quantum speed and the intensity coefficient.

The quantum space above the *Earth* surface has been established by the quantum impact of the elementary evolution, the elementary processes of the minerals of the *Earth*.

The intensity of the quantum system, the *Quantum Membrane,* is loaded by the quantum impact of *gravitation*.

2B3
The drive of the quantum load of *gravitation* above the *Earth* surface is: $IQ_g = \frac{c^2}{\varepsilon_g} = \frac{c^2}{1}$

Other quantum signals can also impact the *quantum system of gravitation*.
Gravitation has its double function:

1. The *approaching* function, the sphere symmetrical expanding acceleration has been generated by the expansion of the *Earth*. The expansion is the consequence of the infinite *blue shift* conflict of the *plasma*. The *approaching* function is a mechanical impact.
    The *blue shift* impact of the approaching (mechanical) function has been proven by the Pound-Sneider-Rebka experiment.
    Earth is expanding and the *Earth* surface is moving, as *plasma* is turning, in the long run, into *Hydrogen* process.

2B6
The expansion is an electron process function as all drives are coming from the electron processes, with the time system of $dt_i = \dfrac{dt}{\sqrt{1 - \dfrac{i^2}{c_{Earth}^2}}}$

As result of the elementary evolution, the speed of the quantum communication on the *Earth* surface is $c_{Earth} = 299{,}792$ km/sec.

Earth surface is in expanding motion with $i = \lim a\Delta t = c_{Earth}$ speed value, corresponding to an electron process function.

2. The *impacting* function of *gravitation* is the anti-electron process *blue shift* quantum impact of the elementary processes of the elementary evolution from the *plasma* to the *Hydrogen* process. The quantum impact is coming from the anti-electron process *blue shift* surplus of the elementary processes with neutron process dominance.

    Anti-electron processes all have the same *IQ* drive (reference to 2A3 and 2A4).

    The *plasma* is of infinite high intensity and infinite high quantum speed. The cooling by *gravitation* slows down the quantum speed and the intensities of elementary processes. The quantum speed of elementary processes is decreasing step by step from the *plasma* state towards the *Earth* surface and is reaching $c_{Earth} = 299,792$ km/sec. As a result of the cooling process, the intensity of the proton process (the expansion) is reaching the intensity of the neutron process (the collapse) and goes it over.

    All those elementary processes with naturally less quantum speed value than c_{Earth} and with proton process dominance are in gaseous state. The *blue shift* impact of the *Earth* surface is affecting these elementary processes; generates *blue shift* conflict and the conflict increases their quantum speed. This impact of the *Earth* is the one establishing the atmosphere.

Ref.
2A3
2A4

Both functions of *gravitation* have a kind of relation to the *blue shift* impacts: In the case of the approaching function, however, the *blue shift* is the proof of the mechanical effect. In the case of the *impacting* function, the *blue shift* itself is the form of the quantum impact.

With reference to the impacting function of gravitation, the atmosphere above the *Earth* is *not equal* to the quantum system of elementary processes.

Quantum system space-times are part of the elementary existence wherever elementary processes in our space-time are to be found. Either they are within the core/mantle of the *Earth* or are part of the atmosphere.

The quantum impact of the anti-electron process *blue shift* surplus is the driving force of *Earth* expansion – *gravitation*.

Minerals are the results of the quantum communication of elementary processes within the mantle of the *Earth* in different regions. Quantum communication usually involves proton process dominant and neutron process dominant elementary processes resulting in infinite format of their appearance.

Our world is about processes, quantum systems, quantum speed and quantum communication, "all with capitals".

2.1
The quantum impulse

S.
2.1

Before dealing with the *quantum impulse,* the meaning of **quantum entropy** must be clarified.

The intensity of the start of the elementary process is: $\dfrac{dmc^2}{dt_o}$

2C1

<u>Proton process:</u> $\dfrac{dmc^2}{dt_o}\left(1 - \sqrt{1 - \dfrac{i^2}{c^2}}\right) = \dfrac{dmc^2}{dt_o} - \dfrac{dmc^2}{dt_i};$ where $i = \lim a\Delta t = c$

2C2

The intensity of the sphere symmetrical expanding acceleration continuous to the time count of $dt_i = \dfrac{dt_o}{\sqrt{1 - \dfrac{i^2}{c^2}}}$ dt_o is corresponding to the time system of the inflexion.

2C3

2C4 Neutron process:
$$\frac{dmc^2}{dt_o}\sqrt{1-\frac{(c-i)^2}{c^2}}\left(\sqrt{1-\frac{i^2}{c^2}}-1\right)=\frac{dmc^2}{dt_i}\sqrt{1-\frac{(c-i)^2}{c^2}}-\frac{dmc^2}{dt_o}\sqrt{1-\frac{(c-i)^2}{c^2}}=$$

$$=\frac{dmc^2}{dt_o}\sqrt{1-\frac{i^2}{c^2}}\sqrt{1-\frac{(c-i)^2}{c^2}}-\frac{dmc^2}{dt_o\sqrt{1-\frac{i^2}{c^2}}}\sqrt{1-\frac{i^2}{c^2}}\sqrt{1-\frac{(c-i)^2}{c^2}}$$

The collapse from the fully expanded state to the status of the inflexion.

2C5 The fully expanded status a kind of inflexion. Collapse can only start if all intensity of the expansion has been expired.
$$dt_{fe}=\frac{dt_i}{\sqrt{1-\frac{(c-i)^2}{c^2}}}\qquad \lim(c-i)=0$$

The full expansion process is acceleration at constant speed, the expiration of the energy intensity of the expansion process against the quantum membrane.

2C6 This is the electron process, the *blue shift* drive of the collapse
$$e_{drive}=\frac{dmc^2}{dt_i}-\frac{dmc^2}{dt_{fe}}=\frac{dmc^2}{dt_i}-\frac{dmc^2}{dt_i}\sqrt{1-\frac{(c-i)^2}{c^2}}$$

Expansion accounts for the internal energy/mass intensity of the process.
Collapse cannot happen without external energy/mass intensity support.
(Categories energy and mass in our *space-time* follow the appearance of the process.)
The principal points of the intensity value of the full expansion are:
-   the quantum system is of infinite low intensity;
-   while the electron process *blue shift* drive is of infinite low value, it cannot be zero; zero as such does not exist; *inflexion* is the status of relative rest.

2C7
$$\frac{dmc^2}{dt_i}\neq\frac{dmc^2}{dt_i}\sqrt{1-\frac{(c-i)^2}{c^2}}$$

The neutron process starts with reference to 2C5 from dt_{fe} (the status of full expansion) indeed. The elementary balance equation in absolute values is:

2C8
$$\frac{dmc^2_x}{dt_p\varepsilon_p}\left(1-\sqrt{1-\frac{i^2}{c^2}}\right)=\xi\frac{dmc^2_x}{dt_n\varepsilon_n}\sqrt{1-\frac{(c-i)^2}{c^2}}\left(1-\sqrt{1-\frac{i^2}{c^2}}\right)$$

2C9 And the electron process *blue shift* drive is:
$$\frac{dmc^2_x}{dt_i\varepsilon_x}\left(1-\sqrt{1-\frac{(c-i)^2}{c^2}}\right)$$

With reference to time counts dt_p and dt_n of the proton and neutron processes in 2C8 versus dt_o in 2C2 and 2C4, - dt_o in these equations is for general meaning. As the neutron-proton process goes through the *inflexion point*, the time definition should correspond to the definition of relative rest (of a system). Absolute rest as such does not exist.

dt_p and dt_n mean the elementary balance with certain intensities of the change, ε_x is the intensity relation of the proton and neutron processes;

dt_i the time system of the electron process, is universal and ensures quantum communication, c_x is the speed of quantum communication, the result of the elementary progress from the *plasma* state and the cooling process of *plasma – gravitation*.

There are two consequences of the balance in 2C8:

1./   2./

$$\frac{dt_n}{dt_p} = \xi \frac{\varepsilon_p}{\varepsilon_n} \sqrt{1 - \frac{(c-i)^2}{c^2}} \; ; \text{ and } \quad \frac{\Delta t_n}{\Delta t_p} = \varepsilon_e \xi \; ; \quad \lim_{i=\lim a\Delta t=c} \xi = \lim \frac{dt_n \varepsilon_n}{dt_p \varepsilon_p} \frac{1}{\sqrt{1 - \frac{(c-i)^2}{c^2}}} = 1$$

2D1

approaching it from above!

There is a residual *quantum energy* intensity "reserve" – the *entropy* consequence within the process (within the mass/energy transformation - in conventional terms), which is equivalent to the intensity difference of the proton and neutron processes: $\Delta q = e_p - e_n$ 2D2

The electron process is the drive of the collapse. Therefore, the electron process as balance component is represented by the neutron process.

The generating intensity of the expansion (of the proton process) covers the intensity need of the drive (the electron process) of the collapse (the neutron process). The cover is permanent, but with the remaining acting *quantum entropy* intensity value.

The intensity of the proton process cover, with reference to 2C2 is: $e_p = \frac{dmc^2}{dt_o}\left(1 - \sqrt{1 - \frac{i^2}{c^2}}\right)$ Ref 2C2

While the cover intensity need of the neutron collapse, with reference to 2C4 is: $e_p = \frac{dmc^2}{dt_o}\sqrt{1 - \frac{(c-i)^2}{c^2}}\left(1 - \sqrt{1 - \frac{i^2}{c^2}}\right)$ Ref 2C4

The *quantum system* has been established by **quantum impulses**, the "instrument" of the generation of the *entropy* of elementary processes.

Entropy is representing a mass/energy intensity capacity of infinite low intensity.

With reference to 2C6 the basis for *entropy* generation is the elementary function of

$$\frac{dmc^2}{dt_p \varepsilon_p}\left(1 - \sqrt{1 - \frac{i^2}{c^2}}\right) = \xi \frac{dmc^2}{dt_n \varepsilon_n}\sqrt{1 - \frac{(c-i)^2}{c^2}}\left(1 - \sqrt{1 - \frac{i^2}{c^2}}\right) ; \quad \text{where } \lim \xi \leq 1$$

Ref. 2C6

The balance in 2C6 relates to the full cycle including the anti-processes.

The *inflexion*, with reference to Section 1, is one and the same; the cycle is one and the same. There are no separate cycles and separate *inflexions* for the processes and anti-processes. The space-time of the elementary process contains both.

The diagrams and the images of the elementary cycles are presented in the book in order to visualise the process. And this might be misleading. Quark and anti-quark processes, as functions of the elementary cycles, the direct and the anti-processes run in parallel rather than in consecutive cycles. The generation of quantum impacts represents both. The anti-processes ensure the full harmony of the *inflexions*. (As it is explained in the next section.)   Ref. 2.2.1

The cycle is about expansion and collapse.

The source of *the quantum impulse* is the missing drive. The electron process *blue shift* impact cannot end with zero impact. "*Nothing*" as such does not exist. Therefore, the neutron collapse always starts from in fact not fully ended electron process status.

In the case of perfect balance, the drive would be: With *entropy* generation, the drive is:

$$e = \frac{dmc^2}{dt_i \varepsilon_e}\left(1 - \sqrt{1 - \frac{(c-i)^2}{c^2}}\right) \qquad\qquad e_\xi = \zeta \frac{dmc^2}{dt_i \varepsilon_e}\left(1 - \sqrt{1 - \frac{(c-i)^2}{c^2}}\right)$$

2D3
2D4

The missing electron process *blue shift* drive is:

2D5
$$\lim \Delta_{quantum} = (1 - \zeta) \frac{dmc^2}{dt_i \varepsilon_e} \left(1 - \sqrt{1 - \frac{(c-i)^2}{c^2}} \right) = 0 ; \qquad \text{where } \lim_{0 \to 1} \zeta = 1$$

As $\lim(1 - \zeta) = 0$

2D6   the speed of the quantum communication of the impulse is: $\lim c_q = \lim c\sqrt{1 - \zeta} = 0$

The missing electron process *blue shift* drive is the intensity of the *quantum impulse*!
= *energy quantum* or *mass/energy quantum*.

2D7   The time system of the infinite low quantum speed corresponds to $dt_i = \dfrac{dt_o}{\sqrt{1 - \dfrac{i^2}{c^2}}}$
But the expression under the square root in this case gives
indeterminate value, as $\lim i = c = 0$

The infinite low value quantum speed of the quantum impulse establishes the space-time of
the quantum impulse as of being infinite small, with infinite long duration – independently
of the value of the intensity coefficient of the elementary process of its generation.
(Division or multiplication of infinite low values results equally in infinite low value.)

In this way, all generating quantum impulses are the results of similar missing infinite
drive values and establish uniform quantum system of equal quantum intensity impact.
The *quantum impulse* is an elementary *blue shift* impact of infinite low intensity.
The generated quantum impulse (*quantum*) is of infinite low *IQ* quantum drive.

The difference in the value of the intensity coefficient does not impact the homogeneity of
the established quantum system: the intensities of all quantum impulses are of infinite low
value. *Quantum* does not take any quantum impact.
Quantum impulse is the end product of electron process. Electron process is impacting or
having been *in blue shift* conflict.
The quantum speed in the quantum system of the *Earth* surface is generated by *gravitation*
and is equal to $c = 299,792$ km/sec.

All elementary processes have their own *space-times* in the quantum system.
For example, there is a space-time belonging to the *Uranium* element with quantum speed
378,163 km/sec, to the *Potassium* elementary processes with 308,335 km/sec.
The quantum speed of the *space-time* is not the integrated value of the quantum impulse.
Quantum speed is the operating speed of the quantum communication within the quantum
system, represented by the electron process *blue shift* impact of elementary processes.
*Quantum impulse*s (quantum) are not about "moving" quantum impacts.
There is a difference between
- quantum impacts initiated by electron process *blue shifts* and
- *quantum impulse* impacts, the result of *entropy* generation.

➤ *Energy quantum* (quantum impulses) are not in conflict – even having in parallel
   different quantum speed and intensity values at their generation.
➤ The infinite low quantum drive remains infinite low value independently of the original
   values of the quantum speed and the intensity of their generation.
Space-times exist in parallel. They transfer all *blue shift* impacts without any modification.

Gravitation is *blue shift* impact and, as such, is in conflict with the electron process *blue
shift* impact and surplus of the elementary processes of the atmosphere (mainly *Oxygen*
and *Nitrogen*, but, in principle, *Helium* and *Hydrogen* processes as well). With the growth

of the height above the *Earth* surface the conflict makes the *blue shift* impact of *gravitation* less and less. With the disappearance of the atmosphere the conflict also disappears.

Blue shift conflicts are barriers for the free propagation of elementary *blue shift* impacts. Conflicts eat off the intensity (the energy) of the transferred quantum signals. Therefore, the transfer of *blue shift* quantum signals needs constant energy intensity supply. The higher the intensity (frequency) is, the higher is the loss. Technical quantum signals propagating within the quantum system have *blue shift* conflict with the electron process *blue shift* impact (surplus) of the elementary processes of the atmosphere, mainly with the *Oxygen* and *Nitrogen* processes.

The energy need is: $E = h\nu$ 2D8

Planck's constant: $h = 6.62606957(29) \times 10^{-34}$ $Joule \cdot sec$; ν is the frequency of the impact. *Blue shift* conflict generates heat, which destroys the quantum impacts and increases intensity.

The *quantum impulse* (energy quantum) is the least possible quantum impact.

There is no difference whether the generating electron process *blue shift* was impact of high or low value *IQ* drive. There is no difference whether the source of quantum impulse generation was the *Uranium* process or the *Helium* process.

The quantum system is the global presence of quantum impulses of infinite low intensity, infinite low quantum speed and infinite small space-time (infinite long duration).

Elementary processes create their own space-time by the speed value of quantum communication. The intensity corresponds to the quantum speed value.

Quantum systems with equal *quantum impulses* of infinite low intensity equally transfer *blue shift* impact of any quantum speed and intensity. Quantum systems operate by the intensity of the impacting electron process.

Summarising:
1. The electron process of the elementary process is *blue shift* quantum impact with certain speed value and intensity. The origin of the *blue shift* impact is the expanding electron process at constant speed.
2. The *quantum impulse* (energy quantum) is the residual impacting capacity of the electron process. It is of infinite low *IQ* value.
3. The quantum impulses (of infinite small space-time) are the ones establishing the basis for quantum systems. The quantum system is the system of quantum impulses impacted by electron process *blue shift* impact of certain quantum speed.
4. Quantum impulses are transferring impacts without any modification. The *blue shift* impact of the electron process is propagating
   - either without any loss of the intensity, if the quantum system is without other impacts;
   - or with the conflicting, breaking impact of other electron process *blue shift* impacts; *blue shift* conflict either stops the propagation or results in the loss of the quantum speed value and intensity.
5. The *quantum* of quantum systems do not impact or influence the intensity of the electron process *blue shift* impacts and transfer all kinds of intensity.

It is important to note that process and anti-process run in parallel and the generation of the quantum impulse relate to the cycle as a single impulse. There are no separate quantum impulses of the direct and the anti-processes.

S.

2.2

2.2

The harmony of the elementary process

The processes and the anti-processes go in parallel, the neutron/anti-neutron and the anti-proton/proton *inflexions* happen simultaneously in line with the intensity characteristics of the elementary process. From the *plasma* state – to the *Hydrogen* process the elementary evolution results in elementary processes with decreasing quantum drive (*IQ*) within the mantle/core of the *Earth*.

The proton and the neutron processes and the anti-neutron and the anti-proton processes mean quantum impulses with $0 \le v \le i$.

2D9
$$\frac{dmc^2}{dt_p}\left(1-\sqrt{1-\frac{v^2}{c^2}}\right); \text{ and } \frac{dmc^2}{dt_n}\sqrt{1-\frac{(c-i)^2}{c^2}}\left(1-\sqrt{1-\frac{v^2}{c^2}}\right);$$

The collapse, which is driven, always needs matter (as *Max Planck* would say). This "matter" is the quantum impact, provided by the proton and the anti-neutron processes.
The quantum impact is the intensity (the measured mass value) of the quark processes:
The (t-c-u) chain means the change of the intensity
>from *top*=180 GeV via *charm*=(1.0 -1.6) GeV to *up*=(2-8) MeV

The (d-s-b) chain means the change
>from *down*=(5-15) MeV via *strange*=(0.1-0.3) GeV to *bottom*=(4-18) GeV

The intensities of the anti-processes, the ($\bar{b}$ - $\bar{s}$ - $\bar{d}$) and the ($\bar{u}$ - $\bar{c}$ - $\bar{t}$) quark chains are corresponding to the same energy/mass (intensity) values just in reciprocal sequence.
[For easier marking the anti-quarks are written in the book, just in (b-s-d) and (u-c-t) forms.]

In the majority of the elementary processes the intensity of the neutron/anti-neutron process is higher than the intensity of the anti-proton/proton process. The intensity relation of the elementary processes depends on the intensity of the electron process. This intensity difference in no circumstances means that the neutron/anti-neutron and the anti-proton/proton *inflexions* are in different time points!
They are in full harmony. The controlling function of the anti-processes is the guaranty:
-   the intensity of the (d-s-b) process is equal to the intensity of the (b-s-d) process; and
-   the intensity of the (u-c-t) process is equal to the intensity of the (t-c-u) process;
    (keeping always in mind the entropy and the quantum impulse corrections).
The neutron process is neutron process, because the second (d-s-b) collapse is missing its (t-c-u) cover. The proton process provides it, which has the extra (t-c-u) chain.
In the anti-direction the anti-proton process is the one, which is missing the ($\bar{b}$ - $\bar{s}$ - $\bar{d}$) cover to its ($\bar{u}$ - $\bar{c}$ - $\bar{t}$) collapse. It is provided by the anti-neutron process.
The proton and the neutron processes keep the internal balance between their other (d-s-b) collapse and (t-c-u) cover quark processes. In the anti-directions the internal balance is obviously ensured between the ($\bar{b}$ - $\bar{s}$ - $\bar{d}$) and the ($\bar{u}$ - $\bar{c}$ - $\bar{t}$) quark processes.
This way with reference to Diag.2.1 and Diag.2.2, the elementary process has 3 *inflexions*. 1-1 is representing the internal balance and the third is result of the internal elementary quantum communication. Each of the three *inflexions* includes the (d-s-b)/($\bar{b}$ - $\bar{s}$ - $\bar{d}$) and the ($\bar{u}$ - $\bar{c}$ - $\bar{t}$)/(t-c-u) quark lines, the change of the direction between the direct and the anti-direct processes. Processes and anti-processes go in parallel.

The homogeneity of the elementary processes is proven by:
- the equal and uniform intensity of the quark processes of all elementary processes;   Ref.
- the fact, that all proton process intensities are uniform and   1G4
  equal and the intensity of the neutron process depends on the   1G5
  characteristics of the elementary process, the value of ε_x, the   2A4
  intensity coefficient;

$$c_x^2 \cdot \varepsilon_x = \frac{c_x^2}{\varepsilon_{x-}} = const$$

S.
2.2.1

2.2.1. Neutron process dominant elementary process are with anti-electron process surplus

S. 2.2.1

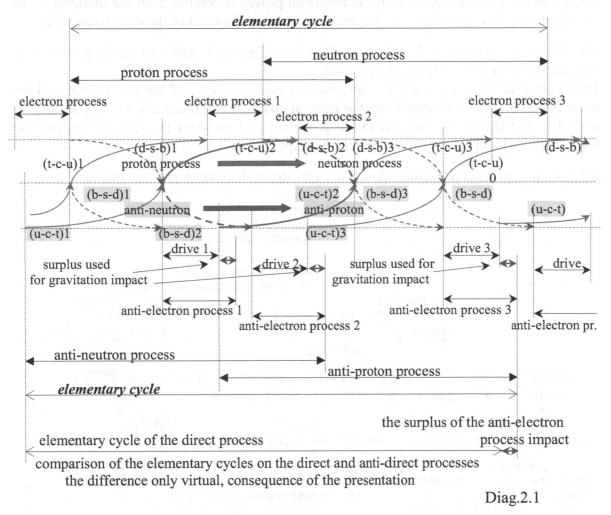

comparison of the elementary cycles on the direct and anti-direct processes
the difference only virtual, consequence of the presentation

Diag.2.1

Diag. 2.1

	The (t-c-u)2 quark process means the proton process		The (d-s-b)2 quark process means it is the neutron process
	The (u-c-t)2 quark process is the anti-proton process		The (b-s-d)2 quark process means it is the anti-neutron process

The diagrams represent the elementary processes in time. The longer is the line, the less is the intensity of the process.

The intensities of the electron processes of the direct and the anti-direct processes are different. The generation of the anti-electron process is more intensive as the neutron process is with increased intensity. But the anti-proton collapse needs a quantum drive of less intensity. Therefore the less intensity in more volume generates surplus.

The anti-process is controlling the direct process.

The intensity values of the process and the anti-process are reciprocal. The anti-electron process of less intensity, taken acting for equal time period with the electron process, results in less driving absolute impact, with still remaining surplus. This surplus is formulating the quantum impact of *gravitation*. [The intensity of the anti-electron process is represented by the whole length of the anti-electron process (= the drive + the delta of the surplus).]

There is no difference in the lengths of the elementary cycles on the process and the anti-process sides. The difference, as the comparison proves is coming from the marking of the anti-electron process surplus – as it is really out (for gravitation) of the elementary cycle.

The elementary cycle has 3 *inflexions*. These 3 *inflexions* include all quark processes in both the direct and the anti-direct processes.

The overlap of the proton and the neutron processes and also of the anti-proton and the anti-neutron processes demonstrates the internal quantum communication between the (t-c-u)/proton and the (d-s-b)/neutron quark processes and between the -(b-s-d)/anti-neutron and the -(u-c-t)/anti-process quark processes respectively.

S.
2.2.2   *2.2.2. Proton process dominant elementary process* with electron process surplus

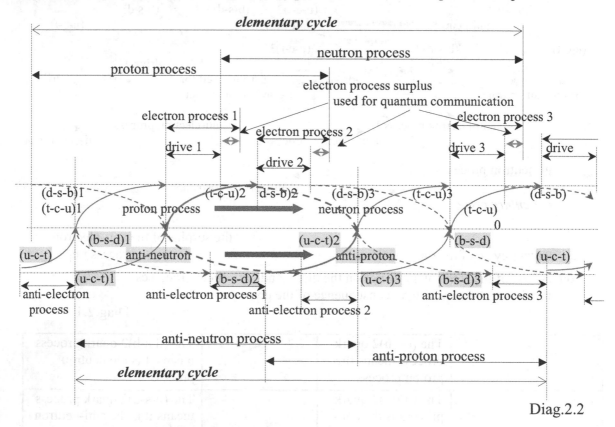

Diag.
2.2   Diag.2.2

The general characteristics of the processes are the same as on the previous diagram.

In the case of the proton dominant elementary processes the electron process surplus is formulating on the direct process side.

The two diagrams prove: the intensities of the electron/anti-electron processes are the ones making the difference between the elementary processes.

$\varepsilon_{ep} > 1$ the intensity of the electron process of the proton process dominant elementary   2D10
      processes is: electron process intensity = the drive + the surplus

$\varepsilon_{en} < 1$ and
                                    $\varepsilon_{en-} = \dfrac{1}{\varepsilon_{en}}$ in the case of neutron process dominant elementary   2D11
$\varepsilon_{en-} > 1$ processes the electron process intensity surplus is
                                                          generating on the anti-process side

The elementary cycle includes 3 electron processes and 3 anti-electron processes.

The internal balance of the proton and the neutron and also the anti-neutron and the anti-proton processes use 2-2 of these electron process drives.

The remaining 1-1 on the direct and the anti-process sides are the drives of the quantum communication between the (t-c-u)/proton and the (d-s-b)/neutron and between the (b-s-d)/anti-neutron and the (u-c-t)/anti-proton processes respectively.

The harmony of elementary processes is unique and standard.

With the constant intensities of the 6 quark processes, the sphere symmetrical expanding acceleration of the *Earth*, the elementary evolution from *plasma* to the *Hydrogen* process produces infinite variety of elementary processes and structures.

The two end stages, the *plasma* and the *Hydrogen* processes have been connected.

Otherwise both are with missing components. The *Hydrogen* process is missing the (d-s-b) collapse; the *plasma* is missing the (t-c-u) expansion. The final collapse of the *Hydrogen* process – at the stage of the infinite high intensity of its quantum membrane, having all elementary processes in *Hydrogen* process status – generates *plasma*. The evolution of *plasma* results in *Hydrogen* process.

The *Hydrogen* process is accumulating as (t-c-u) and collapsing as (d-s-b). The *inflexion* produces (b-s-d), the *plasma* and the elementary evolution goes on as (u-c-t).

Quantum impulse is generating at each of the three *inflexions* of the elementary processes.

Elementary processes have three elementary quark lines in balanced communication.

There is internal quark process balance separately within the proton and within the neutron processes as well. The overlap in the quark process balance between the proton and the neutron processes means the internal elementary communication. This overlap is also open for the elementary communication with other elementary processes.

The elementary communication is easy, as

- the intensity of the drive, the electron process is always the one, which is establishing the intensity of the neutron process;
  therefore the neutron process can be driven by the electron process and covered by the proton process of any elementary process in elementary communication.
- the intensities of all proton processes are equal and being kept constant by the anti-proton processes, at the same time they are relativistic, as the intensity is representing the proton/neutron process relation.
- the anti-processes follow the direct elementary line; and

The quantum impulse is the missing electron process *blue shift* drive of the elementary cycle including the direct and the anti-processes. (The difference between the integrated intensity of the proton/anti-proton and the anti-neutron/neutron processes.)

In proton process dominant elementary processes the *quantum impulse* is generating on the anti-direction side (as the direct process is with electron process *blue shift* surplus);

For the same reason, in the case of neutron process dominant elementary processes, the *quantum impulse* is formulating on the direct elementary side.

2.3
The quantum system of the *Earth*

Quantum systems are with infinite number of quantum impulses, the results of the remains of electron process *blue shift* drives. This means quantum systems only exist as products of elementary processes.

Ref.
2D5

The integrated intensity of the quantum system with infinite number of equal quantum impulses (quantum) of infinite low intensity (as of infinite long duration) is infinite low. With reference to 2D5 the intensity of the quantum impulse can be written as

2E1

$$\lim q = \lim(1-\zeta)\frac{dmc^2}{dt_i\varepsilon}\left(1-\sqrt{1-\frac{(c-i)^2}{c^2}}\right) = 0 \ ;$$

Quantum systems are communicating if they are impacted!

The quantum system above the *Earth* surface is the result of the elementary evolution from the *plasma* to the *Hydrogen* process and has been induced by the *blue shift* impact of the sphere symmetrical expanding acceleration of the *Earth – gravitation*. The quantum impact of *gravitation* is establishing the quantum speed on the *Earth* surface. The quantum impacts within our quantum system (space-time) propagate by the speed of the quantum communication of the *Earth*.

This is not about flying photons. Quantum (quantum impulses) transfer quantum impacts by the speed of quantum communication.

Quantum signals with more quantum speed value than our quantum system are capable of entering into and can impact our quantum system. All those of less intensity value cannot

Ref.
S.1.7

and, being impacted, are speeded up as the *Hydrogen, Oxygen* and the other elementary process of the atmosphere. With reference to Section 1.7 the speed of quantum communication is the main characteristic of the space-time.

Quantum impacts with higher quantum speed propagate within our quantum system as well, but their increased intensity is expiring step by step. This is not about working against the quantum impulses (*quantum*). This is rather having conflict with and working against the *blue shift* impacts of *gravitation*.

The quantum signal of the *Sun* is entering as light impact and conflicting as heat.

Quantum impacts with quantum speed and intensity higher than our system (like the electron process *blue shift* impact of the *Sun*) generate *blue shift* conflict within the space-time of the *Earth*.

This follows from the electron process formula:

2E2

$$\frac{dmc_S^2}{dt_i\varepsilon_S}\left(1-\sqrt{1-\frac{(c_S-i_S)^2}{c_S^2}}\right) = n\frac{dmc_E^2}{dt_i\varepsilon_E}\left(1-\sqrt{1-\frac{(c_E-i_E)^2}{c_E^2}}\right) \ ; \quad \frac{c_S^2}{\varepsilon_S} = n\frac{c_E^2}{\varepsilon_E} \ ; \quad \text{or} \quad IQ_S = n\cdot IQ_E$$

The impact of the *Sun* corresponds to high number of quantum impacts within the quantum system of the *Earth* atmosphere generating *blue shift* conflict and resulting in heat.

Single electron process *blue shift* impacts, even of higher intensity, cannot increase the intensity of the whole quantum membrane of the *Earth*. Those either establish their own space-time in parallel with or within our quantum system (if having sufficient feed) or the impact simply will be lost.

The *blue shift* impact of *gravitation* above the *Earth* surface is general and overwhelming. Establishing its own quantum system in parallel within our system means partial impact of the quantum impulses of our space-time. Because of the loss through conflict at the boundaries for keeping this local space-time acting it needs constant intensity supply.

A single separated light signal within a vacuum chamber meets the overall quantum impact of *gravitation*, whatever is the quantum speed of its generating intensity. The speed of the quantum communication corresponds to the speed value of the electron process *blue shift* impact of *gravitation* within the chamber. The *blue shift* conflict eats off the speed difference and establishes the speed of the propagation of the light signal equal to the resulting speed value.

All quantum impacts and signals have been transferred by the *energy quantum* of the quantum systems without any change. In the case of the light signal above in 2E2, the conflict is the one which is modifying the speed value.

Ref.
2E2

Quantum impacts generated within our space-time on the *Earth* surface cannot impact quantum system with higher intensity.

Elementary processes in gaseous state on the *Earth* surface have the *blue shift* conflict which keeps them in gaseous status.

Elementary processes within the *Earth* core have their own space-time system and quantum speed of communication.

2.4

Space-times, difference in time counts

S.
2.4

Quantum impulses (energy *quantum*) equally represent $\lim IQ_q = 0$.

The electron process drives the neutron process to *inflexion*. The anti-neutron process generates the anti-electron process with sufficient *blue shift* impact to collapse the anti-proton for starting a new cycle. The case is controlled by the anti-process: for any change in the elementary process, first the anti-process has to be modified.

The anti-proton process is neutral in the same way as the neutron process is.

If the anti-proton process cannot be driven to collapse, the missing *inflexion* cannot initiate a new cycle whatever is the reason.

It is important to underline: while the generation of the quantum impulse is $\lim IQ_q = 0$,

the quantum drive may cover a variety of intensity coefficient values – quantum impulses are uniform and equal. Energy *quantum,* quantum impulses of quantum systems, therefore, equally transform quantum signals and electron process *blue shift* impacts of any *IQ* drive.

Ref.
S.1.7

Space and time, with reference to Section1.7, depends on the value of the quantum speed.

If we take an event happening in parallel within two *space-times* x and y with different speed values of quantum communication c_x and c_y;

and mark the event itself by v;

the durations of the event within the two space-times will be:

For space-time x:	For space-time y:	In the case of $c_x > c_y$ and $\varepsilon_x < \varepsilon_y$
$$dt_x = \dfrac{dt_{xo}}{\sqrt{1-\dfrac{v^2}{c_x^2}}}$$	$$dt_y = \dfrac{dt_{yo}}{\sqrt{1-\dfrac{v^2}{c_y^2}}}$$	the result is obvious: the gradient of the time flow in quantum system x is less $= dt_x < dt_y$ – the duration of the event is shorter.

2F1

2F2

Note: ε the intensity coefficient is reciprocal to the real intensity.
As the intensity value itself is reciprocal to the duration, the intensity
coefficient is in direct relation with the time gradient.

The explanations in our *Earth* conditions mean:

1. The quantum impulses (energy *quantum*) of the quantum system above the *Earth* surface are equally transferring all impacting quantum signals, independently of their quantum speed and intensity.

2. Quantum signals with higher *IQ* quantum drive – cannot be directly detected in our space-time on the *Earth*. They have been detected through their conflicting impact!

Ref. S.5.2

3. With reference to 2F1, a quantum signal, with quantum speed of $c_x = 383,869$ km/sec, the quantum speed of the Giza *pyramid* is taken as an example, acting in parallel within our space-time on the *Earth*, with $c_E = 299,792$ km/sec quantum speed.

4. The event represents a quantum impact to our space-time generated by the *pyramid*: The time system of relative rest is taken for $dt_{Eo} = dt_{xo} = 1$.

v representing the event	$\dfrac{dt_x}{dt_E}$ the time count relation	In the case of events of low v speed values, (in usual, everyday cases), while the two time systems are different and always with less time count within the system of reference of the pyramid, the difference in practical terms has no significance.
1.00	0.99999999999783	
10.00	0.999999999783	
100.00	0.9999999783	
1,000.00	0.99999783	
10,000.00	0.999783	
100,000.00	0.97644	
200,000.00	0.8727	With the increase of the speed the difference
250,000.00	0.727	at $\lim v = c_E$ is maximal and growing.
293,792.999	**0.00013**	

Table 2.1

Table 2.1

2F3

The electron process quantum signal of the pyramid generates a certain conflicting quantum membrane, since the quantum communication indicates $n = 1.64$ times more *blue shift* impact within the space-time of the *Earth*.

$$n = \frac{IQ_x}{IQ_E} = \frac{c_x^2}{c_E^2} = 1.64$$

$$\varepsilon_E = \varepsilon_x = 1 \text{ as taken}$$

The quantum speed values of minerals within the *Earth* are increased.

Space-times of elementary processes within the *Earth* mantle transfer the *blue shift* impacts by their own original quantum speed value. *Blue shift* conflict cannot be avoided. The acting conflict is the driving force of the sphere symmetrical expanding acceleration. This conflict generates heat and the cooling is reducing the conflict and the quantum speed values.

The speed of the expanding acceleration of and at the surface is $i_E = \lim g\Delta t = c_E$.

This cannot be differently, since the resulting by the elementary conflicts reduced speed value itself is the speed value at the surface, the speed of the expanding acceleration.

Might two space-times with different quantum speed values exist in parallel?

It is better to pose this question in a different way: Might other space-times exist in parallel within our quantum system space-time on the surface of the *Earth*?

The answer is *yes* and *no*.

No = if the *IQ* value of the space-time is less than the intensity of the *Earth*. The example is the case where the elementary processes are in gaseous state. They become gaseous, because their space-time has been modified by the increase of their quantum speed.

Yes = if the *IQ* value is higher than the *IQ* of the *Earth* system and the impact is constant always compensating the lost intensity due to the conflict with *gravitation*.

The space-time with quantum speed c_x and with intensity coefficient ε_x generates *blue shift* impact to the overall quantum system:

$$\frac{dmc_x^2}{dt_i\varepsilon_x}\left(1-\sqrt{1-\frac{(c_x-i_x)^2}{c_x^2}}\right) \quad \text{with} \quad IQ_x = \frac{c_x^2}{\varepsilon_x} \qquad \text{2G1}$$

This *IQ* value will be lost constantly
to the conflict with *gravitation* by a quantum drive equal to $\quad IQ_{Earth} = \dfrac{c_E^2}{\varepsilon_E} = \dfrac{c_E^2}{1}$ \qquad 2G2

For the constant compensation of the loss, the source shall be a capacity of:	As the delta intensity of the loss shall be equal zero:	
$IQ_x = nIQ_E$		$IQ_x - nIQ_E = 0$

2G3

In the case where any impact the impacting space-time itself cannot be detected within the time system of the *Earth*, it is because its time system is of different rate. (The case is almost similar to the radio signals just radiofrequencies are generated at the same, equal quantum speed of the *Earth* quantum system and the detection is within this same space-time.) Here, the difference is in the quantum speed value, which means different space-time.

In the case of an electron process *blue shift* quantum impact of a space-time,
quantum speed of $c_x = 383,689$ km/sec; and intensity coefficient of $\varepsilon_x = 1.00421$

Ref. 2F3

the time system of the impacting space-time in line with 2F3 is 1.63 times more intensive than the time system of the *Earth* with $c_E = 299,792$ km/sec and $\varepsilon_E = 1$.

Ref. 2F1 2F2

In the case of similar events v within a space-time with c_x quantum speed and within the system of reference of the *Earth* with $c_E = c_y$, with reference to 2F1 and 2F2 the two time counts are different.

Similar events happen shorter in time within space-times with higher quantum speed than the quantum speed on the *Earth* surface.

S. 2.4.1

2.4.1. The value of the speed of the quantum communication on the *Earth* surface has been established by *gravitation*

The quantum impact of *gravitation* provides constant load to the quantum system above the *Earth* surface. The quantum speed of this impact – as the result of the elementary evolution from *plasma* to the *Hydrogen* process – is $c_{Earth} = 299,792$ km/sec.

The speed value of the sphere symmetrical expanding acceleration of the *Earth* is $i = \lim a \cdot \Delta t = c_{Earth}$ and this is the approaching, mechanical impact of *gravitation*.

The quantum membrane above the *Earth*, in this way, is impacted by c_{Earth} quantum speed.
As presented in Section 5, if the intensity of this quantum membrane for a certain reason has been changed, the quantum speed value also would be changed.

Ref. S.5

S. 2.3.2

2.4.2. The impact of the *Earth* space-time on elementary processes

The space-time of the *Earth* belongs to $c_E = 299,792$ km/sec quantum speed of communication. Elementary processes have their own specific quantum speed value.

From the point of view of an elementary process, there is no significance concerning what is the time count on the surface of the *Earth*. This time count is based on the rotation and the circulation of the *Earth* around the *Sun*. This is, indeed, an excellent practical tool for describing processes on the *Earth* surface, but it has in fact nothing to do with the duration, the intensity and space-time of elementary processes.

The value of the quantum speed is the only one having impact on the space-time of elementary and other processes.

Ref. Table 1.1

With reference to Table 1.1, all elementary processes of the periodic table have space-times and intensity values different than the space-time and the intensity on the surface of the *Earth*; except for the *Hydrogen, Helium, Oxygen* and *Nitrogen* processes, which have been driven above the surface by the quantum drive of the *Earth*.

In this way, in the case of an event, which could be considered as common and the same for all elementary processes, the time counts within each of the elementary processes –

Ref. S.1.3 S.2 2G1

measured even by the time count of the *Earth* – would be different as the intensities of their space-times are different.

With reference to Section 1.3 and Section 2: $c_x^2 \cdot \varepsilon_x = const$ - the higher the quantum speed is, the higher is the intensity (the less is the value of the intensity coefficient).

As the intensity of the processes is reciprocal to the time count; and ε_x the intensity coefficient as indicator itself is reciprocal to the real intensity and its value is proportional to the duration of the event, comparing the duration of the same event

2G2

in the space-time of the elementary process of x and y – if $\varepsilon_x < \varepsilon_y$ it gives $\Delta t_x < \Delta t_y$

Meaning: the time count in the space-time of the elementary process x is less.

In other words, the elementary process x has used less of its internal capacities for "managing" the event.

There is no importance where the two x and y elementary processes have happened. It is not the placing which decides what the internal time counts of the elementary processes are. Both can equally be placed on the surface of the *Earth*, still their own internal time system would count the time-flow of the event (as long as their quantum speed value has not been changed to the quantum speed of the *Earth*).

Elementary processes have their own quantum speed in their mineral status.

They communicate with each other within the *Earth* for having the best optimal elementary status, but keeping their own quantum speed.

Melting out elementary processes of their mineral status means:
- creating electron process *blue shift* conflict by external electron process *blue shift* impact (heating);
- the melted status is of common quantum speed value, the result of the heating *blue shift* conflict: (*E* means external impact: on the *Earth*)

2G3

$$\frac{dmc_x^2}{dt_i\varepsilon_x}\left(1-\sqrt{1-\frac{(c_x-i_x)}{c_x^2}}\right) + n\frac{dmc_E^2}{dt_i\varepsilon_E}\left(1-\sqrt{1-\frac{(c_E-i_E)}{c_E^2}}\right) = \frac{dmc_z^2}{dt_i\varepsilon_x}\left(1-\sqrt{1-\frac{(c_z-i_z)}{c_z^2}}\right)$$

- the conflict makes elementary processes fluent;
- the fluency at different conflict level – meaning at different temperatures – gives the chance to separate the elementary components, as the quantum speed and the intensity values of the minerals are different;

- the cooling means taking off the *blue shift* conflict, thus causing the hardening of the fluent elementary process(es);
- the hardening happens at the quantum speed of the external impact of the *Earth*, as the melted mix is with the intensity dominance of the external impact;
- in this way, the "clean" elementary processes on the surface of the *Earth* becomes acting with the quantum speed of the *Earth*.

Coming back to the main message of the elementary processes on the *Earth* surface:

The internal duration of the event, within the elementary process can be counted by the time system of the *Earth* (as we do not have any other). But it will still be characterised by the internal time-flow of the elementary process (meaning, it can be less and more).

2.4.3. Time impact of the quantum membrane of increased intensity

The increased intensity does not change the intensity coefficient, but increases the speed value, therefore, the same event happens for shorter time period, it becomes of more intensity:

$$\Delta t_x = \frac{\Delta t_o}{\sqrt{1 - \dfrac{v^2}{c_x^2}}}$$

The higher is c_x the shorter is Δt_x ; v is the same in both cases.

S. 2.3.3

2H1

Δt_o means the same status of relative rest in both cases, otherwise the comparison loses its common basis.

The time-count on the *Earth* surface represents the space-time of the *Earth*.

It can be used as the time-count technique for characterising elementary events, but it shall be clear that this is nothing other than the one which the processes are compared to.

All elementary processes have their own space-time, function of their quantum speed value.

2.4.4. The space-time of the *Earth* surface

The space-time on *Earth* has been established by *gravitation*.

The quantum impulses are from the elementary processes; the *blue shift* impact is from the sphere symmetrical expanding acceleration, establishing the speed and the space-time.

The space-time of quantum impulses, with reference to 2D6, is infinite small. The space-time of the *Earth* contains infinite number of quantum impulse space-times.

S. 2.3.4

Ref. 2D6

The space-time of the quantum impulse (*quantum*) is the smallest; with duration of infinite length; existing in infinite numbers. Quantum impulses cannot take impacts; they are the tools of the transfer of the quantum impacts of the elementary quantum communication of quantum systems space-times. The communication is based on the electron process *blue shift* impact of elementary processes, which is of the same time system.

Events happen in the space-time of quantum impulses for infinite duration.

The higher the quantum speed of the space-time is, the shorter is the internal time count of the same event within.

Plasma as space-time is of infinite intensity and infinite short internal time-count.

The *Hydrogen* process is at the other end with the infinite length of the event.

We, on the *Earth*, are somewhere closer to the infinite length, while the time-count within almost all elementary processes of the periodic table (exempt the eight with proton process dominance) and beyond is shorter. And the process itself is of infinite short/infinite length in time.

What is the difference between the *Hydrogen* process and the quantum impulse?
- The quantum drive of the *Hydrogen* process is of infinite low intensity, but the electron process is generated by the proton process. Its quantum speed can be increased.
- The quantum impulses (quantum) have been generated by the elementary processes and the infinite low quantum speed cannot be increased.

S.
2.4

2.5
The harmony with *h*, the *Planck* constant and the *Planck* formula

The *quantum impulse* (the *quantum*) and the *quantum impact* are different categories.
The quantum impulse is, with reference to 2D5, the missing from the elementary cycle electron process *blue shift* drive:

Ref.
2D5
$$\lim q = \lim \Delta_{quantum} = (1 - \zeta) \frac{dmc_x^2}{dt_i \varepsilon_x} \left(1 - \sqrt{1 - \frac{(c_x - i_x)^2}{c_x^2}} \right) = 0; \quad \text{where } \lim_{0 \to 1} \zeta = 1$$

As $\lim(1 - \zeta) = 0$ and $\lim IQ_q = 0$, the intensity of the quantum impulse is independent on the type of the elementary process and it is always equal to infinite low value.

The <u>quantum impact or quantum signal</u> is the electron process *blue shift* impact itself:

2I1
$$e_x = \frac{dmc_x^2}{dt_i \varepsilon_x} \left(1 - \sqrt{1 - \frac{(c_x - i_x)^2}{c_x^2}} \right)$$

Within our *space-time* on the surface of the *Earth* the value of the quantum speed is constant and equals to $c_x = c_E = 299,792$ km/sec, but the intensity of the electron process can be impacted by different technical methods, mainly by acceleration and conflict. The intervention does not change the elementary process, but certainly modifies its appearance. Therefore 2I1 can be written as:

2I2
$$e_{impact} = \frac{dmc_E^2}{dt_i \varepsilon_x} \left(1 - \sqrt{1 - \frac{(c_E - i_E)^2}{c_E^2}} \right) = \frac{n_x}{\varepsilon_x} \frac{dmc_E^2}{dt_i} \left(1 - \sqrt{1 - \frac{(c_E - i_E)^2}{c_E^2}} \right);$$

or in absolute terms within the time system of the Earth:

2I3
$$E_{impact} = \frac{n}{\varepsilon_x} \frac{dmc_E^2}{dt_i \varepsilon_E} \left(1 - \sqrt{1 - \frac{(c_E - i_E)^2}{c_E^2}} \right) = \frac{n_x}{dt_i \varepsilon_x} dmc_E^2 \Delta t_E \left(1 - \sqrt{1 - \frac{(c_E - i_E)^2}{c_E^2}} \right); \quad \text{where}$$

2I4
Ref.
S.
2.1

$$h = \Delta mc_E^2 \Delta t_E \left(1 - \sqrt{1 - \frac{(c_E - i_E)^2}{c_E^2}} \right);$$
Calculating with unit mass, this formula gives a constant value on the surface of the *Earth*.

It was measured by *Max Planck* and its value the is *Planck's constant*:
$$h = 6.62606957(29) \times 10^{-34} \text{ Joule} \cdot \text{sec}.$$
As result of the technical *intervention* the impact of the elementary

2I5
$$v = \frac{n_E}{\Delta t_E \varepsilon_x} = \frac{n_x}{dt_i \varepsilon_x}$$

parameters dt_i and n_x (the number or the intensity of the impacting electron processes) is appearing within the circumstances of the *Earth* obviously in different dimension.
ε_x is dimensionless and is without change

The formula in 2I5 is the measured frequency on the *Earth*.
The intensity of the electron process *blue shift* impact in 2I1 gives the *Planck formula*:

2I6
The energy/work need of the quantum impact in absolute terms is: $E = h \cdot v$

3
Process/anti-process relations

Process and anti-process relations are the basics of relativity!

If the elementary balance of a change/motion or event is assessed from the point of view of a certain *system of reference*, the elementary balance shall also be assessed from the other direction, from the point of view of the *opposite elementary cycle* as well.
Only in this case can we reach valid results and discuss the real consequences.
Therefore, assessing events, the elementary processes together with their anti-processes as well is not only justified but also obligatory.

The key objective of the anti-process is the elementary balance and, in this way, the assurance of the continuity of the elementary cycle.
Nature is change, *inflexion*, about generation of *quantum entropy* and *quantum impulse*, about the fluency of processes: collapse continues with expansion, expansion is followed by collapse.
The diagram below with explanation refers to Diag. 1.1 in Section 1.

Ref Diag. 1.1

Diag.3.1

Diag. 3.1

The change in absolute terms means the balance: $\varepsilon_p \neq \varepsilon_n$ as $dt_p \neq dt_n$ **3A1**

$$\frac{dmc^2}{dt_p\varepsilon_p}\left(1-\sqrt{1-\frac{v^2}{c^2}}\right) = \xi\frac{dmc^2}{dt_n\varepsilon_n}\sqrt{1-\frac{(c-i)^2}{c^2}}\left(1-\sqrt{1-\frac{v^2}{c^2}}\right)$$ the intensities of the processes as basic parameters are different **3A2**

Proton and neutron processes are events-accelerations!

dt_p and dt_n	denote the status of the rest of the elementary process – the *inflexions* of the proton and the neutron processes, which are together with the speed of the quantum communication, the characteristics of the elementary process.	**3A3**

The time systems of the elementary *inflexions* (the rest status) are not absolute *zero* values.
These are the time systems of the space-time of overall space-time matrix.

$$dt_v = \frac{(dt_p = dt_n = dt_o)}{\sqrt{1-\frac{(a\Delta t)^2}{c^2}}}$$ dt_v, the time count within the elementary system in motion is about *constant change*. Its actual value depends on the actual speed value, which is function of the time and the acceleration; function of the intensity of the process. **3A4**

> Δt is the measurement within the actual dt_x time system.
> c the quantum speed of the elementary process is *constant*. This varies between values $\lim c = 0$ and $\lim c = \infty$. The one end is plasma and the other is the *Hydrogen* process.

The definition of the *space-time matrix* above means the system of space-times, built on each other without end.

The time count between the rest status $v = 0$ (the *inflexion*) and $v = i_x = \lim c_x$ (the electron process stage) depends on the intensity of the process.

The proton process as expansion (transformation of mass status into energy) and the neutron process as collapse (retransformation of energy into mass) go through all speed values and the corresponding time systems between the two endpoints. Mass and energy statuses are relative categories; their only meaning is the certain direction of the change.

The only way to describe the elementary process is by ε the intensity of the processes.

The intensities of the proton and the neutron processes are different, but these are the main characteristics of the elementary cycle, the basis of the drive and the intensity of the electron process.

Therefore, dt_v the time definition of the change in both directions indeed corresponds to the intensity of the processes.

[The quantum speed values for all elementary processes are different, but are equal and constant to the proton, the neutron and the electron processes of the elements. The constant change during the electron process is the acceleration at constant speed value.]

3A5   The number of *inflexions* for the proton and neutron processes is equal. This means: $p = n$

3A6   In intensity terms – in the real occurrence of events – however, this means: $\dfrac{dp}{dt_p} \neq \dfrac{dn}{dt_n}$,

which would exclude common *inflexions*, since, in this case, the process times are different.

The *harmony* of the process can only be guaranteed if:

3A7
$$\frac{dp}{dt_p}\Delta t_p + \frac{dn}{dt_n}\Delta t_n = \left| -\frac{dn}{dt_n}\Delta t_n - \frac{dp}{dt_p}\Delta t_p \right|; \quad \text{and} \quad \frac{dp}{dt_p}\Delta t_p = \frac{dn}{dt_n}\Delta t_n \quad \text{as} \quad \varepsilon = \frac{1}{\Delta t}$$

Anti-processes are controlling the harmony of the cycle as presented in Diag.3.1.

Explanation: Proton/neutron processes – without anti-process control – would leave
            elementary balance open:

The absolute balance of the change is in order, but the intensity difference between the two opposite directions would remain without "*control*": Different intensities, different time components of the proton and neutron processes would represent different process cycles. Different process cycles without *control* result in different sequence of events, with no *inflexions*.

Inflexion is the key: no inflexion means no elementary process.

Alongside the absolute mass/energy balance of the event, if the expansion and the collapse in intensity terms do not correspond to each other, the continuity is disrupted. The harmony would be over; there is no proton process without neutron process and vice versa without common *inflexion*. The number of proton and neutron processes should be equal.

The proton process as expansion is the physical opposite to the neutron collapse, but they are of different intensities. Their intensity relation depends on the type of the elementary process. The anti-process direction is a physical need for ensuring the continuity and the mass/energy balance!

The time frames, with reference to Diag.3.1 of the elementary expansion: $\Delta t_{exp} = \Delta t_p + \Delta t_{a-n}$ and the elementary collapse: $\Delta t_{coll} = \Delta t_n + \Delta t_{a-p}$	and $\Delta t_p = \Delta t_{a-p}$ and $\Delta t_n = \Delta t_{a-n}$	The balance is in order: $\Delta t_{exp} = \Delta t_{coll}$	3A8

The anti-, and the direct processes run in parallel and ensure the time (and intensity) control of the processes.

Anti-processes do not change the characteristics of elementary process: the intensity relations and the proton and neutron process dominances remain unchanged.

Elementary processes with *proton process dominance*
- the proton process provides electron process *blue shift* drive with surplus;
- but the *blue shift* impact of the electron process is of low intensity value;
- therefore, the neutron process is also of low intensity;
- the intensities of the anti-processes shall correspond to the intensities of the normal (proton and neutron) processes, since only this condition guarantees the harmony, the parallel and symmetrical run of the complete process:

$$\varepsilon_n = \varepsilon_{a-n}; \quad \text{and} \quad \varepsilon_p = \varepsilon_{a-p};$$ 3A9

- anti-neutron process fulfils expanding function and is in control (with the same intensity) of the neutron collapse;
- in line with the equal values of the proton/anti-proton, neutron/anti-neutron intensities as represented in 3A9, the intensities of the electron and anti-electron processes become different, as:

$$\frac{dt_n}{dt_p} = \varepsilon_e = \frac{\varepsilon_p}{\varepsilon_n}\sqrt{1 - \frac{(c-i)^2}{c^2}}; \quad \text{and} \quad \frac{dt_{a-p}}{dt_{a-n}} = \varepsilon_{a-e} = \frac{\varepsilon_{a-n}}{\varepsilon_{a-p}}\sqrt{1 - \frac{(c-i)^2}{c^2}}; \quad \text{and} \quad \varepsilon_{a-e} < \varepsilon_e$$ 3A10

(higher ε value means less electron process intensity, less ε means more)

- the increased intensity of the anti-electron process (as the coefficient is of less value) guarantees the completion of the elementary cycle with higher proton process intensity;
- the increased intensity of the anti-electron process is, in fact, the internal mass/energy intensity reserve of the proton intensity dominant elements;
- the electron process *blue shift* impact of the "normal" direction remains in surplus and ready for "external" elementary use for quantum communication, corresponding in this way to the acting anti-electron process intensity of higher value (the intensity coefficient is less).

Ref. 3A10

Elementary processes with *neutron process dominance*
- the proton process provides electron process *blue shift* impact of increased intensity;
- the intensity of the neutron process is higher than the intensity of the proton process;
- with reference to 3A11 below, the intensity of the *blue shift* impact of the anti-electron process is of decreased value but with surplus;

$$\frac{dt_{a-p}}{dt_{a-n}} = \varepsilon_{a-e} = \frac{\varepsilon_{a-n}}{\varepsilon_{a-p}}\sqrt{1 - \frac{(c-i)^2}{c^2}}; \quad \text{and} \quad \varepsilon_{a-e} > \varepsilon_e; \quad \text{as} \quad \varepsilon_n > \varepsilon_p \text{ and } \varepsilon_{a-n} > \varepsilon_{a-p}$$ 3A11

(higher ε value means less electron process intensity, less ε means more)

- the anti-neutron process with higher intensity than the anti-proton process generates anti-electron process *blue shift* impact of less intensity, but with surplus;
- anti-electron process *blue shift* surplus not used as drive for collapsing anti-protons – as those are of less intensity – creates anti-electron process *blue shift* impact for driving/loading the *gravitation* process;

- anti-electron process *blue shift* surplus is the mass/energy intensity source for the *blue shift* impact of *gravitation;*
- *gravitation* is the acting component of the overall elementary evolution.

Process/anti-process approach is of high importance:
- for the continuity of the elementary energy/mass and mass/energy transformation – the absolute balance is the most important;
- while proton and neutron, anti-proton and anti-neutron processes have different intensities;
- with the benefit of the mass/energy intensity surplus of the process dominancy (either proton process or neutron) is used for elementary evolution (neutron process dominancy) or quantum communication (proton process dominancy);

The intensities of the electron and anti-electron processes

Electron process surplus in normal elementary direction causes surplus and conflict, which explain the gaseous, liquid and other states of elementary processes.
The increased intensities of the electron process *blue shift* impacts of neutron process dominant elementary processes result in specific elementary structure with direct contact through the anti-electron process *blue shift* impact with *gravitation.*

The intensity relations of the electron process drives are:

3B1
$$\varepsilon_e = \frac{\varepsilon_p}{\varepsilon_n}\sqrt{1 - \frac{(c-i)^2}{c^2}}\ ; \quad \text{and} \quad e_e = \frac{dmc^2}{dt_i\varepsilon_e}\left(1 - \sqrt{1 - \frac{(c-i)^2}{c^2}}\right); \quad \begin{array}{l} \text{as} \\[4pt] \varepsilon_p = \varepsilon_{a-p} \end{array}$$

$$\varepsilon_{a-e} = \frac{\varepsilon_{a-n}}{\varepsilon_{a-p}}\sqrt{1 - \frac{(c-i)^2}{c^2}}\ ; \quad \text{and} \quad e_{a-e} = \frac{dmc^2}{dt_i\varepsilon_{a-e}}\left(1 - \sqrt{1 - \frac{(c-i)^2}{c^2}}\right); \quad \begin{array}{l} \text{and}\ \varepsilon_n = \varepsilon_{a-n} \\[8pt] \text{and} \\[6pt] \varepsilon_e = \dfrac{1}{\varepsilon_{a-e}} \end{array}$$

if $\varepsilon_e < 1 \longrightarrow \varepsilon_{a-e} > 1$; if $\varepsilon_e > 1 \longrightarrow \varepsilon_{a-e} < 1$;

3B1 simplifies the subject about the intensities of the electron and anti-electron processes:

3B2
3B3
$$n\frac{dmc^2}{dt_i\varepsilon_{a-e}}\left(1 - \sqrt{1 - \frac{(c-i)^2}{c^2}}\right) = \frac{dmc^2}{dt_i\varepsilon_e}\left(1 - \sqrt{1 - \frac{(c-i)^2}{c^2}}\right) \quad \begin{array}{l} \text{3B1 is valid} \\[6pt] \text{if}\ \ \varepsilon_e = \dfrac{\varepsilon_{a-e}}{n}\ \text{and}\ \ \varepsilon_{a-e} = n\cdot\varepsilon_e \end{array}$$

The intensity of the electron process can be written equally for the direct and the anti-processes as:

3B4
$$n\frac{dmc^2}{dt_i\varepsilon_e}\left(1 - \sqrt{1 - \frac{(c-i)^2}{c^2}}\right) = \frac{dmc^2}{dt_i\dfrac{\varepsilon_e}{n}}\left(1 - \sqrt{1 - \frac{(c-i)^2}{c^2}}\right)$$

3B2, 3B3 and 3B4 prove:
If in one direction the intensity is *n*-times more, in the other direction the reduced by *n*-times intensity as surplus has *n*-times more volume.

➢ In elementary processes with *proton process dominance,*
   - the surplus is generating on the direct side – as the proton process is of increased intensity, resulting in electron process *blue shift* surplus, conflict and increased activity – the key for quantum communication;
   - the anti-electron process is of increased intensity generated by the anti-neutron process for driving the anti-proton process of increased intensity.

> In elementary processes with *neutron process dominance*,
  - the surplus is formulating on the anti-direction side and the anti-electron process *blue shift* surplus is establishing the internal mass/energy intensity potential of the elementary process, source of *gravitation*;
  - the electron process is of increased intensity, generated by the proton process in order to provide the intensity of the neutron collapse.

Elementary processes are clear proofs of *relativity*:

The intensity increase on one side is compensated by the surplus of the other side, keeping the relations always in balance.

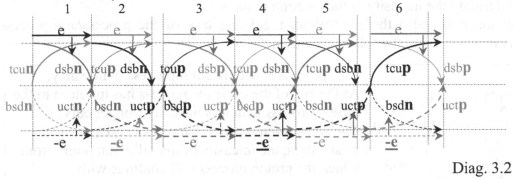

Diag. 3.2

Diag. 3.2

Diag. 3.2 gives more information for understanding the process and the elementary cycle in Diag. 3.1.

In both, in the proton and in the neutron processes as well, are expanding and collapsing functions, represented by the corresponding quark processes. The difference is that there is a *top-charm-up* chain in the proton process without balancing *down-strange-bottom* line and a *down-strange-bottom* chain in the neutron process without balancing *top-charm-up* line. In fact, these not in balance quark processes give the character of and the reason for the communication of the proton and the neutron processes.

For the explanation of the communication the diagram above is divided into 6 sections.

In sections 1, 3 and 4 and 6 the balance of the collapse and the expansion is ensured within the processes themselves. In section 2 and 5 the balance is the result of the communication by the processes in sections 1 and 3 and 4 and 6.

Sections 1 and 4 provide the collapsing function and sections 3 and 6 add the expanding one. In this way, section 2 and 5 are the meeting points for collapse and expansion. Two collapses with an expansion are the neutron processes and two expansions with a single collapse are the proton process.

As it is proven, there are three lines of quark processes in elementary communication. These three lines are presented in the diagram by three different colours.

Each elementary process consists of these three lines. The elementary communication of these three lines establishes the elementary process itself.

As the segments are demonstrating, one of the lines is in collapse, the other is in expansion and the third is the drive of the collapse. It cannot be differently, since the one expanding or collapsing cannot be functioning as the drive of the collapse at the same time as well.

The process is simplified in the explanations and the other diagrams of this book, as

tcu**p** is representing the proton process;

dsb**n** is representing the neutron process;

bsd**n** is the anti-neutron process (as expanding function);

uct**p** is the anti-proton process (as collapse);

e marks the electron process and –*e* the anti-electron process.

The process in sections 2 and 3 is a complete proton process, communicating with the neutron process in section 2. The process in 4 and 5 is a complete neutron process, which is communication with a proton process in section 5. Sections 1 and 6 are presenting the half of a complete neutron and proton processes.

The length of the curves in the figure depends on the number of the cycles of the elementary processes. In the *Hydrogen* process it is only a single cycle, with never ending open neutron process. *Plasma* representing infinite number of elementary processes.

The number of the proton, neutron and electron processes is always equal.

There is a difference between the intensities of the proton and the neutron processes. This difference is defining the intensity of the electron process.

Ref.
Diag.
3.2

Diag. 3.2 demonstrates also the controlling, regulating task of the process/anti-process relation. This is presented in the following diagrams.

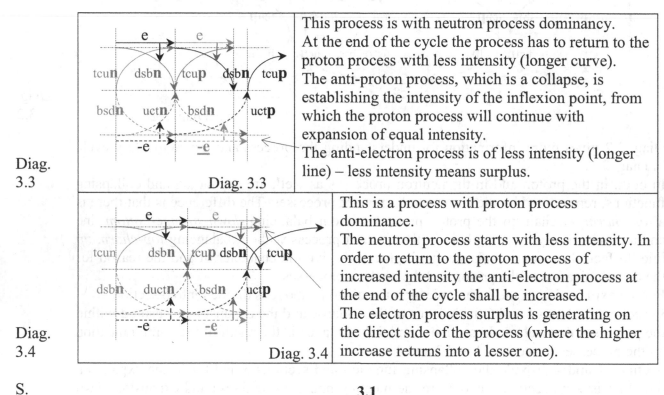

Diag. 3.3 — Diag. 3.3	This process is with neutron process dominancy. At the end of the cycle the process has to return to the proton process with less intensity (longer curve). The anti-proton process, which is a collapse, is establishing the intensity of the inflexion point, from which the proton process will continue with expansion of equal intensity. The anti-electron process is of less intensity (longer line) – less intensity means surplus.
Diag. 3.4 — Diag. 3.4	This is a process with proton process dominance. The neutron process starts with less intensity. In order to return to the proton process of increased intensity the anti-electron process at the end of the cycle shall be increased. The electron process surplus is generating on the direct side of the process (where the higher increase returns into a lesser one).

S.
3.1

3.1
Proton process dominance

The *blue shift* impact intensity demand of the neutron process of elementary processes with proton process dominance is decreased, as the neutron process is of less intensity:

3C1

$$\frac{d(process)}{dt_{neutron}} < \frac{d(process)}{dt_{proton}} \quad \text{if} \quad dt_n > dt_p$$

The proton process is more intensive, meaning: while the time count in the neutron system is Δt_n, in the proton system it is Δt_p.

Electron process is the elementary link between the proton and neutron processes. Continuity is about balanced *inflexion*, which, in other words, also means: the *equality* of the numbers of the proton and neutron processes.

3C2

Electron
process
is at stage

$$dt_i = \frac{dt_o}{\sqrt{1 - \frac{i_x^2}{c_x^2}}}$$

If the *generation* of electrons as end product of the proton process is more intensive than their use as starting point of the neutron process, the intensity surplus is obvious.

The intensity need of the collapse initiates the intensity of the drive and the cover accordingly. Without anti-processes, the control of the cycle, the difference in the intensities, could result in the destruction of the process: The difference in the durations of the expansion and the collapse in this case could destroy the continuity of the elementary process. The anti-processes control the elementary process and close the cycle.

The intensity of the electron process is: | The intensity of the anti-electron process is:

$$e_e = \frac{dmc^2}{dt_i \varepsilon_e}\left(1 - \sqrt{1 - \frac{(c-i)^2}{c^2}}\right) \qquad e_{a-e} = \frac{dmc^2}{dt_i \varepsilon_{a-e}}\left(1 - \sqrt{1 - \frac{(c-i)^2}{c^2}}\right) \qquad 3C3$$

While $\varepsilon_p = \varepsilon_{a-p}$ and $\varepsilon_n = \varepsilon_{a-n}$; $\varepsilon_p > \varepsilon_n$ as it is about proton process dominancy:

$$\frac{\Delta t_n}{\Delta t_p} = \varepsilon_e = \frac{\varepsilon_p}{\varepsilon_n}\sqrt{1 - \frac{(c-i)^2}{c^2}} \; ; \text{ and } \frac{\Delta t_{a-p}}{\Delta t_{a-n}} = \varepsilon_{a-e} = \frac{\varepsilon_{a-n}}{\varepsilon_{a-p}}\sqrt{1 - \frac{(c-i)^2}{c^2}} \; ; \text{ and } \varepsilon_e > \varepsilon_{a-e} \qquad 3C4$$

If we denote $\varepsilon_{a-e} = \dfrac{\varepsilon_e}{y}$, in line with 3C4 above $y > 1$ and substituting it into 3C3:

$$e_e = \frac{dmc^2}{dt_i \varepsilon_e}\left(1 - \sqrt{1 - \frac{(c-i)^2}{c^2}}\right) \quad \begin{array}{l}\text{The intensity} \\ \text{equivalent of the anti-} \\ \text{electron process is:}\end{array} \quad e_{a-e} = \frac{dmc^2}{dt_i \frac{\varepsilon_e}{y}}\left(1 - \sqrt{1 - \frac{(c-i)^2}{c^2}}\right)$$

3C5
Ref
3B3
3C6

electron process action time is: $\Delta t_n = \Delta t_{a-n}$ as $\varepsilon_n = \varepsilon_{a-n}$ 3C7

anti-electron process action time is: $\Delta t_{a-p} = \Delta t_p$ as $\varepsilon_p = \varepsilon_{a-p}$ 3C8
    3C9

Proton process dominancy means: $\varepsilon_p > \varepsilon_n$: $\Delta t_p < \Delta t_n$; and $\Delta t_{a-p} < \Delta t_{a-n}$

The difference in the durations also means different *blue shift* impacts in time:

If $\dfrac{\Delta t_n}{\Delta t_p} = z$ and $\Delta t_p = \dfrac{\Delta t_n}{z}$ and $z > 1$;   3C10

$$E_e = e_e \cdot \Delta t_n \text{ and } E_{a-e} = e_{a-e} \cdot \Delta t_{a-p} \quad (\text{as } \Delta t_{a-p} = \Delta t_p)$$

$$E_e = \frac{dmc^2}{dt_i \varepsilon_e}\left(1 - \sqrt{1 - \frac{(c-i)^2}{c^2}}\right)\Delta t_n \; ; \quad \text{and } E_{a-e} = \frac{dmc^2}{dt_i \frac{\varepsilon_e}{y}}\left(1 - \sqrt{1 - \frac{(c-i)^2}{c^2}}\right)\frac{\Delta t_n}{z} \; ; \qquad 3C11$$

$$\text{as } \Delta t_{a-p} = \frac{\Delta t_{a-n}}{z} \text{ and } \Delta t_n = \Delta t_{a-n} \qquad 3C12$$

Coefficients z and y in the equations above shall be equal as both characterise the same intensity difference between the proton and neutron processes.

As $z = y$ in absolute terms the equality and the balance is proven: $E_e = E_{a-e}$

The equality in 3C7 above can also be proven as intensities and time durations are reciprocal $\varepsilon_e = \dfrac{\Delta t_n}{\Delta t_p} = \dfrac{\Delta t_{a-n}}{\Delta t_{a-p}} = z = y$ and $\varepsilon_{a-e} = \dfrac{1}{z} = \dfrac{1}{y}$ 3C13

From 3C4 it follows that, if the number of participating electron processes within the normal process direction were N, result of N end-stages of the proton process, the number of anti-electron process *blue shift* impacts is   as $z > 1$ $\dfrac{N}{z} < N$ 3C14

The electron process intensity surplus within the normal process direction is: $s = N - \dfrac{N}{z}$ 3C15

This is the quantum communication capacity of proton process dominant elements.

S.
3.2.

3.2
Neutron process dominance

The proton process in neutron process dominant elementary processes happens for longer duration. The neutron process is more intensive and, therefore, it takes less time.

$$\frac{d(process)}{dt_{proton}} < \frac{d(process)}{dt_{neutron}};$$

The neutron process is more intensive:

While the time count in the neutron system is Δt_n,

3D1   if $dt_n < dt_p$ in the proton system it is Δt_p.

The deduction is similar to the proton process dominancy case in the previous section, with the difference that the anti-electron process surplus is generating on the anti-process side.

In the case of the proton process dominance, the electron process *blue shift* surplus creates *blue shift* conflict and quantum communication capacity.

In the case of neutron process dominance, the generating anti-electron process *blue shift* surplus is representing the internal energy of the process and is the drive of the elementary evolution and provides quantum impact to *gravitation*.

Ref.
3C10-
3C15

With reference to 3C10-3C15, the elementary processes with neutron process dominance give:

3D2 $\varepsilon_e = \dfrac{\varepsilon_{a-e}}{y}$; with $y > 1$ and $\dfrac{\Delta t_{a-p}}{\Delta t_{a-n}} = z$ and $\Delta t_{a-n} = \dfrac{\Delta t_{a-p}}{z}$ and $z > 1$

Ref.
3C15

The absolute balance is guaranteed, as with reference to 3C15: $E_e = E_{a-e}$ and $z = y$, as

3D3
3D4

$$E_e = \frac{dmc^2}{dt_i \frac{\varepsilon_{a-e}}{y}}\left(1 - \sqrt{1 - \frac{(c-i)^2}{c^2}}\right)\frac{\Delta t_{a-p}}{z}; \quad \text{and} \quad E_{a-e} = \frac{dmc^2}{dt_i \varepsilon_{a-e}}\left(1 - \sqrt{1 - \frac{(c-i)^2}{c^2}}\right)\Delta t_{a-p}$$

Ref.
S.1
S.2

and the electron process *blue shift* surplus (as anti-electron process impact) is indeed generating on the anti-process side.

S.3.1 and S.3.2 give the proof of the reciprocal intensity relation of the electron and anti-electron processes.

The lesson of the above sections with proton and neutron process dominances is that
- while the absolute balance is perfectly guaranteed, quantum communication happens in line with the intensity values of the electron and anti-electron processes:
  ➤ as intensity values provide constant and online balance at integrated level:

3D5
$$\frac{dmc_x^2}{dt_i}\left(1 - \sqrt{1 - \frac{(c_x - i_x)^2}{c_x^2}}\right)$$

for reaching $\dfrac{dmc_x^2}{dt_p}\left(1 - \sqrt{1 - \dfrac{i_x^2}{c_x^2}}\right) = \xi \dfrac{dmc_x^2}{dt_n}\sqrt{1 - \dfrac{(c_x - i_x^2)}{c_x}}\left(1 - \sqrt{1 - \dfrac{i_x^2}{c_x^2}}\right)$

Neutron processes are neutral, therefore, the balancing part to the electron process drive and the proton process cover in the 3D5 equation can be from any elementary relations. The intensity potential is acting and the balance is building up in the elementary quantum communication.

This is the feature of elementary processes with electron process intensity surplus.

➤ in anti-direction, with reference to 3A12, the intensity surplus in the anti-electron process side is used for the expansion of the *plasma*, for building up *gravitation*, for the generation of elementary processes with decreasing neutron process intensity;

the work of the *quantum membrane* is logically using all available sources in line with their occurrence, with reference to Section 1.2 and 1F5

Ref. 3A12

$$e_{a-e} \rightarrow [e_{a-e\downarrow} + e_g] \rightarrow e_{a-e\downarrow} = (n - \Delta n)\frac{dmc_x^2}{dt_i \varepsilon_{-m}}\left(1 - \sqrt{1 - \frac{(c_x - i_x)^2}{c_x^2}}\right) + [e_g]$$

3D6

$[e_g]$ is with the Δn anti-electron process impact on *gravitation*

This is characteristic of the elementary processes with anti-electron process intensity surplus.

- the continuity of the processes is the guarantee for the absolute balance.

3.3
Elementary communication

S. 3.3

Proton and also neutron process dominances are the drives of elementary evolution.

Both are with mass/energy capacity, initiating the progress, the elementary change from *plasma* to *Hydrogen*. The tools of the utilisation of the drive, however, are different:

- proton process dominant elementary processes stream to communicate to utilise their electron process *blue shift* surplus establishing communication and balance with other elementary processes;
- neutron process dominant elementary processes are the ones communicating with proton process dominant elementary processes, utilising the available electron process surplus and proton cover provided;
- neutron process dominant elementary processes also utilise the surplus of their anti-electron process *blue shift* impact for developing and feeding *gravitation*.

Diag.3.5 on the next page demonstrates the communication of a proton process dominant elementary process (A) with elementary process (B) with neutron process dominancy.

Ref. Diag. 3.5

"A" with proton process of increased intensity and with electron process *blue shift* surplus drives and provides cover to the neutron process of "B" of increased intensity.

"B" is with anti-neutron process of increased intensity (as corresponding to the intensity of the neutron process), and with anti-electron process *blue shift* surplus (as response) drives and covers the anti-proton process of "A" with increased intensity.

The objective of the elementary communication of these elementary processes with intensities of opposite directions is the establishment of the optimal 1/1 balance state.

The intensities of both directions have been utilised.

Diag.3.5 helps to understand the dynamics of the quark processes as well.

Two parallel quark processes within the proton/neutron and anti-neutron/anti-proton processes, the *top-charm-up* and *down-strange-bottom* quark process lines are the basis for internal mass/energy relations. The *third*, in the proton process of "A" the extra *top-charm-up* quark process line is the one providing electron process *blue shift* drive and proton process cover to the extra *down-strange-bottom* quark process line of the neutron dominant "B"; while the *bottom-strange-down* quark process line of the anti-neutron process of "B" provides anti-electron process *blue shift* drive and anti-neutron process cover to the *up-charm-top* quark process line of the anti-proton process of the elementary process of "A".

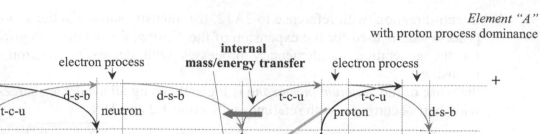

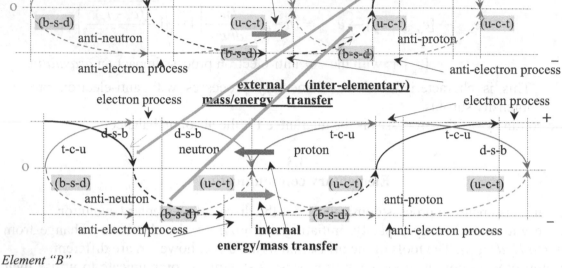

Diag.
3.5

Diag. 3.5

In this particular case in Diag.3.5 the electron process *blue shift* surplus and the proton process cover are provided by elementary process "A" to the neutron process of "B"; and

the anti-electron process *blue shift* and the anti-neutron process of elementary process "B"

drives and covers the anti-proton process of "A".

Elementary communication utilises the electron process *blue shift* surplus of the direct and the anti-processes. Both decrease the quantum impact potential of elementary processes, the contribution to *gravitation* and to elementary communication.

Elementary processes with neutron process dominance are with elementary anti-electron process potential.

> This is the potential which feeds the expansion-*gravitation* within the *Earth* structure.
> Minerals with neutron process dominant elementary processes out of the *Earth* mantle - while usually having been in certain composition with proton process dominant elementary processes (like oxides, carbons, nitrates, calcites, silicates, etc.), which are balancing their anti-electron process surplus - still possess certain intensity potential. The explanation is that the elementary communication reduces the anti-electron process potential, but the value of the quantum speed, which is formulating as consequence, remains above the quantum speed of the *Earth*. The melting out of the elementary process from the mineral composition means *blue shift* conflict, resulting as cool down product quantum communication with the quantum speed of the *Earth*.

Ref.
Diag.
3.5

The elementary processes in Diag.3.5 are in fact cycles with two *inflexions* at the two ends of the quark processes. Therefore, the *entropy* generation belongs to the cycles as introduced in Section 2.

3.3.1. The intensity of the electron process in elementary communication

With the exception of the known *Hydrogen, Helium, Carbon, Nitrogen, Oxygen, Silicon, Sulphur* and *Calcium* elementary processes all proton process intensities are less than the intensity of the neutron process. This raises the question: How can be an electron process capable to drive a neutron process, if the intensity of the neutron process is more than the intensity of the proton process? How can generate a proton process an electron process with increased intensity?

Answering the question there are facts to be taken into account:

1. Elementary process relations are in intensities. The electron process is connecting intensities.

    The elementary balance in absolute terms is in order.

2. The neutron process is the driven one and it is neutral in principle, but the anti-neutron process is the one, which is establishing the intensity value demand and this value is the one, which shall be met by the drive of the electron process.

3. The *IQ* of the electron process and its proportions are:

$$IQ = n\frac{c^2}{\varepsilon}$$

- If $\varepsilon < 1$; means $n < 1 \rightarrow$ the drive is only partially meeting the intensity demand, any external drive is welcome for reaching the intensity balance status.

$$IQ = \frac{c^2}{\left(\dfrac{\varepsilon}{n}\right)}$$

- If $\varepsilon > 1$; means $n > 1 \rightarrow$ the drive is in surplus, ready for use.

- The *IQ* of the drive in fact in both cases gives identical formats.

    n denotes the proportion of the electron process within the elementary process

4. The process/anti-process relation keeps the elementary evolution ongoing.

    In global terms: $\varepsilon_+ = \dfrac{1}{\varepsilon_-}$ Because of the entropy generation of each elementary cycle however: $\varepsilon_- = \dfrac{1}{\varepsilon_+} - \Delta$

$$\varepsilon_- = \frac{n}{\varepsilon_+} - \eta;$$

and

$$\varepsilon_+^{new} = \frac{n}{\varepsilon_- - \eta}$$

The intensity of the anti-electron process determines the intensity of the proton process of the new cycle, which in both variants results in "higher" intensity:

If $\varepsilon_- > 1$ and $n > 1 \rightarrow \varepsilon_+^{new} > \varepsilon_+$ getting closer to $\varepsilon_+ = 1$;

If $\varepsilon_- < 1$ and $n < 1 \rightarrow \varepsilon_+^{new} > \varepsilon_+$ getting farer from $\varepsilon_+ = 1$.

η means the impact of the quantum impulse of the entropy generation.

5. In the quantum communication of proton process and neutron process dominant elementary processes the intensity of the demand of the drive and the cover is formulating within the elementary process with neutron process dominance, while the drive and the cover is available within the elementary process with proton process dominance.

    The anti-neutron process of the neutron process dominant elementary process is the one which determines the intensity demand of the drive of the neutron process.

    While the relations in the anti-directions are the opposite, there is a huge difference in the elementary processes here: the anti-proton process is the one, which determines the intensity of the proton process of the following elementary cycle!

    The process with reference to Section 3.3 and Diag.3.5 is:

    The electron process of the elementary process with surplus and proton process dominance drives and covers the neutron process of the neutron process dominant elementary process. The anti-electron process of the neutron process dominant elementary process drives and covers the anti-proton process of the proton process dominant elementary process.

But the communication relates only to the quark processes without balance.

The other two quark processes having been in internal balance are without impact.

This way the precise description is:

The electron process surplus of the elementary process with proton process dominance drives the (*d-s-b*) process of the neutron process dominant elementary process and covers it by its (*t-c-u*) chain; while the anti-electron process surplus of the neutron process dominant elementary process drives the *(u-c-t)* process of the proton process dominant elementary process and covers it by its (*b-s-d*) chain.

This way the outstanding from the internal balance elementary quark cycles of both participants become balanced and create their common quark cycle with the participation of the anti-directions and under the controlling function of the anti-direction of the neutron process dominant elementary process.

- the drive is coming from the proton process dominant elementary process;
- the controlling is coming from the neutron process dominant elementary process.

The (*b-s-d*) establishes the intensity demand of the (*d-s-b*) driven by the electron process surplus of the proton process dominant elementary process and the (*u-c-t*) driven by the anti-electron process surplus of the neutron process dominant elementary process determines the intensity of the (*t-c-u*).

- as the intensity of the anti-electron process is reciprocal to the intensity of the electron process and the intensity demand of the (*d-s-b*) is controlled by (*b-s-d*);
- the electron process surplus available drives the neutron process of the neutron process dominant elementary process by the exact intensity value of the demand.

The reciprocal intensity value of the anti-electron process means, the intensity of the drive of (*u-c-t*) corresponds to the intensity of the anti-electron process of the neutron process dominant elementary process, which generates the (*t-c-u*) chain within the elementary process of the proton process dominant elementary process in line with the neutron process dominant elementary process.

Thus the original question/concern this way can be answered:

The electron process available within the elementary processes with proton process dominancy drives the neutron process of the neutron process dominant elementary process in line with the demand and the anti-process of the neutron process elementary process fully controls the communication.

The quantum speed of the communication is the one belonging to the elementary process with proton process dominance (the drive).

The number of the acting elementary drives as below in 3G6 and the generating conflict shall correspond to the *IQ* of the intensity drive demand of the other, neutron process dominant elementary process of the elementary communication.

3E6   The *IQ* of the electron process drive is: $IQ = n_x \dfrac{c_x^2}{\varepsilon_x}$

As result of the communication, the two types of elementary processes compose a stable and balanced elementary structure, with standard elementary control. Both elementary processes of the composition manage their own elementary process with the direct communication of one of their quark process chain segment.

3.4
Hydrogen and the anti-process

The uniqueness of the *Hydrogen* process, also explaining the difficulties of the measurement of its neutron process is that the extra *(d-s-b)* quark chain of the neutron process and
the extra *-(u-c-t)* quark chain of the anti-proton process (the driven one) are both from the *Helium* elementary process.
These two quark process chains have been developing under the *blue shift* impacts of the electron and the anti-electron processes of the *Helium* elementary process.

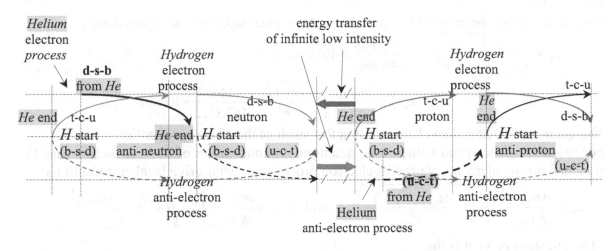

Diag. 3.6

Inflexions always and everywhere mean the evolution of elementary relations.
All elementary relations have extra quark processes which are originating from the expanded status of the previous elementary process, demonstrating the full adequacy with the process.
In the case of the *Hydrogen* process
all *inflexions* are from the *Helium* process and
the extra *(d-s-b)* and *(u-c-t)* quark lines are the continuity of the *Helium* process expansion.

Hydrogen process generates electron process and anti-electron process *blue shift* impacts of infinite low intensity. The neutron and the anti-proton processes of the *Hydrogen* process are driven by external *blue shift*. (Neutrons and, consequently, the anti-protons as well are neutral and follow the drive, independently of whether the drive is an internal one or an external one.)
The neutrons of the *Hydrogen* process have been driven in elementary communication by other elementary processes. The electron process of the *Hydrogen* process remaining in *blue shift* surplus within the communication creates conflict.

In the case of *Hydrogen* generation from other elementary mixes, such as from water, the *Oxygen* is the elementary process replacing *Helium* in relation to the extra *(d-s-b)* and *(u-c-t)* quark chain and the electron and anti-electron processes at the start.

S.
3.5

3.5
Anti-processes and isotopes

Isotopes are elementary processes with damaged balance. Damaged balance state means deviation in process/anti-process relations.

The fission of the *Uranium-235* process is none other than the result of the generated by water flow increased electron process *blue shift* conflict.
The internal balance of the *Uranium-235* process is already with *blue shift* surplus, with the decrease of the intensity of the electron process *blue shift* impact, $\varepsilon_{eU235} = 0.651$ against

$\varepsilon_{eU238} = 0.6375$:

(The surplus either means less intensity value or the original higher one with partial use.)

3F1
$$\varepsilon_{eU238} = \frac{\varepsilon_{pU238}}{\varepsilon_{nU238}}\sqrt{1 - \frac{(c-i)^2}{c^2}} = \frac{92.6688}{145.3602} = 0.6375; \quad \text{and}$$

3F2
$$\varepsilon_{eU235} = \frac{\varepsilon_{pU235}}{\varepsilon_{nU235}}\sqrt{1 - \frac{(c-i)^2}{c^2}} = \frac{92.6688}{142.3312} = 0.6510$$

The quantum speed is also changing. It is the result of the *blue shift* conflict.
For calculating the speed value of the quantum communication of this conflicting case the standard for the elementary process electron process *blue shift* drive is taken for unit value.

3E3
$$e_{U238} = \frac{dmc^2_{U238}}{dt_i\varepsilon_{U238}}\left(1 - \sqrt{1 - \frac{(c_{U238} - i_{U238})^2}{c^2_{U238}}}\right) = 1 \quad \text{and} \quad e_{U238} = \frac{dmc^2_{U235}}{dt_i\varepsilon_{U235}}\left(1 - \sqrt{1 - \frac{(c_{U235} - i_{U235})^2}{c^2_{U235}}}\right) = 1$$

3E4

The question is what is the IQ_{U235} and the c_{U235} quantum speed value giving the same drive impact?

$$\frac{c^2_{U238}}{\varepsilon_{U238}} = \frac{c^2_{U235}}{\varepsilon_{U235}}; \quad \text{and} \quad c_{U235} = c_{U238}\sqrt{\frac{\varepsilon_{U235}}{\varepsilon_{U238}}}$$

(the dt_i time count of the electron process value in the brackets are taken equal for both normal and isotope cases)

[Here the case is different than in the process of the elementary evolution from *plasma* to *Hydrogen* with $c^2_x \cdot \varepsilon_x = const$, where the elementary quantum speed value is decreasing with the intensity decrease of the elementary process.
Here the process remains the same, with damaged elementary (isotope) structure, with the intensities of the components modified.]

The values of the quantum speed for *U-235* and *U-238* in 3E2 demonstrate the conflict: *382,175 km/sec* for *U-235* and *378,173 km/sec* for *U-238*. The intensity of the collapse to be driven becomes less, while the proton process and the *blue shift* impact generation remains the same:
As proof of the approach, the *IQ* value of *U-235* remains the same $2.243 \cdot 10^{11}$ km^2/sec^2, equal to the standard *IQ* value of *Uranium-238*. The less electron process intensity is compensated by the increased quantum speed. This standard *IQ* value means a relative intensity quotient increase for *U-235*.

The reason for all these changes is that the intensity of the expansion (weight) of the proton process in fact remains the same (≈ 92), while the intensity of the collapse (weight) of the neutron process becomes reduced (from ≈ 145.3602 to ≈ 142.3312).
The already existing but still controlled (within *U-235*) *blue shift* conflict becomes increased by the electron process *blue shift* surplus of water. (The *Cadmium* control rods within nuclear reactors increase the balancing capacity of the *U-235* load, as the *Cadmium*

process is communicating with the electron process *blue shift* surplus of the water. With the control rods with the *Cadmium* process out of the zone the balance deviation of the *Uranium-235* process, the result of the additional *blue shift* surplus of the water, is acting immediately.)

The *Uranium-235* process is with electron process *blue shift* surplus (while *U-238* is not).

The *U-235* process is also with anti-electron process *blue shift* surplus, but of decreased value comparing with the natural *U-238* process. The relative increase of the *IQ* drive and the electron process *blue shift* surplus/conflict can be tolerated within the natural elementary structure of the *Uranium* process up to around 5%. If the content is increased and in addition is under the impact of other electron process *blue shift* source (water), the elementary process of the *Uranium* cannot be maintained anymore!

The fission of the *Uranium-235* obviously is for resolving the conflict by producing elementary processes with less *IQ* values. The problem is that the gradient of the decrease of the *IQ* value, the drive of the neutron process, is higher than the gradient of the decrease of ε_x the intensity coefficient of the electron process (ε_x – the coefficient of the relation between the intensities of the proton and neutron processes). As it is demonstrated on Diag.3.7 here below.

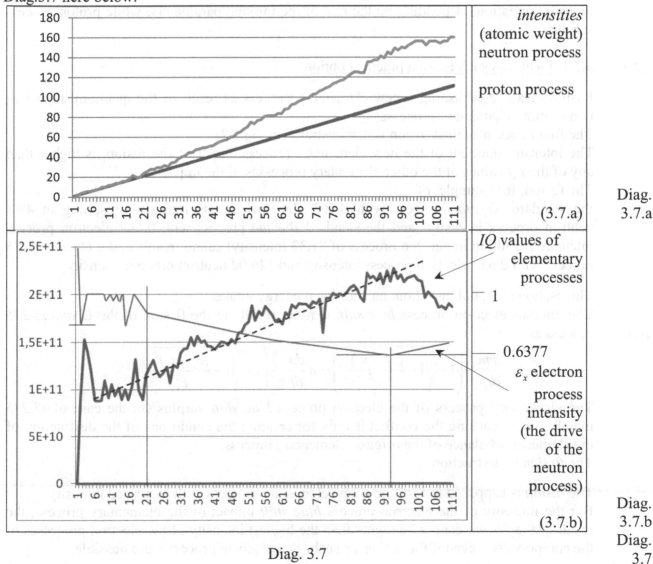

Diag. 3.7

Diag. 3.7.a

Diag. 3.7.b

Diag. 3.7

The difference in the gradients means:
- the intensity of the neutron process to be driven is decreasing by certain gradient,
- but the gradient of the decrease of its drive is about the double of this value (as *IQ*, the intensity of the electron process proves it) – with reference to Diag.3.7.b.

The new elementary processes developing as results of the fission (fission products – isotopes), are in conflicting balance – with missing electron process *blue shift* drive and proton process cover.

3F3 $U_{92}^{235} \rightarrow Kr_{36}^{92} + Ba_{56}^{141}$ As example, the formulating *Krypton* and *Barium* fission products have neutron process intensities of *56.0* and *85.0* respectively,
with increased electron process *blue shift* drive demand, while normal *Krypton* is with 47.8 and normal *Barium* is with 81.33 neutron process results in so-called *alpha* radiation intensities.

Fission-product-isotopes in elementary structures out of mass/energy balance like the above ones further generate new isotopes themselves. There might also be formulating elementary processes this way with proton process dominance, where the even decreased *IQ* value is higher than the one necessary for the balanced state.

Missing electron process *blue shift* drive and proton process cover is causing *beta* or *gamma* "radiation" depending on the rate of the missing parameters, while proton process surplus.

S.
3.5.1   ### 3.5.1. Fusion as such is not a practical option

Fusion means establishing a new elementary process as result of the quantum impact of two or more elementary processes.
The first concern against fusion is demonstrated in Fig.3.7:
The intensity quotient of the new elementary process, result of the fusion, is higher than any of the *IQ* values of the other elementary processes of the impact.
The fusion, for example, of
the standard *Krypton* process with 0.763 electron process intensity (resulting in 47.8 neutron process intensity) and the standard *Barium* process with 0.697 electron process intensity (resulting in neutron process of 81.33 intensity) cannot result in the *Uranium-238* process with 0.637 electron process intensity and 146.02 neutron process intensity.

This is, however, not just about an additional energy intake.
The obvious electron process *blue shift* conflict results in the fission of the *Uranium-235*
3G1   process as:

$$\frac{dmc_U^2}{dt_i\varepsilon_U}\left(1-\sqrt{1-\frac{(c_U-i_U)^2}{c_U^2}}\right) + n\frac{dmc_x^2}{dt_i\varepsilon_x}\left(1-\sqrt{1-\frac{(c_x-i_x)^2}{c_x^2}}\right)$$

The elementary process of the electron process *blue shift* surplus (in the case of *U-235* mainly water) causing the conflict is only for creating the conditions of the destruction of the elementary balance of the original elementary process.
The *fission* is destruction.

The *fusion* is supposed to result in a new elementary process with increased intensity.
But the intensity of the electron process *blue shift* impact of the elementary process, the result of the *fusion,* cannot be higher than the highest intensity of the electron processes of the components – even if the *fusion,* as such, as a practical process were possible.

$$n_a \frac{dmc_a^2}{dt_i \varepsilon_a}\left(1 - \sqrt{1 - \frac{(c_a - i_a)^2}{c_a^2}}\right) + n_b \frac{dmc_b^2}{dt_i \varepsilon_b}\left(1 - \sqrt{1 - \frac{(c_b - i_b)^2}{c_b^2}}\right) + n_x \frac{dmc_x^2}{dt_i \varepsilon_x}\left(1 - \sqrt{1 - \frac{(c_x - i_x)^2}{c_x^2}}\right) \qquad \text{3G2}$$

Processes a and b are the electron process *blue shift* impacts of the elementary processes of the *fusion*. Elementary process x is for the generation of the electron process *blue shift* conflict, for increasing the intensity of the process – in other words, for external energy supply.

Processes a and b, have their elementary balance separately. They cannot compose another new elementary process with higher intensity. They may have elementary conflict, which may result in their separate destruction, never a fusion. The quantum speed, as the result of the conflict and the intensity increase, can be increased reaching a higher common value for all the components, but the intensity coefficient of the elementary processes cannot be changed. To keep these in joint elementary quasi mix would need additional external energy, which, as magnetic force, would create additional electron process impact.

Fusion is not the opposite process to *fission*.
This process in 3G2 is also destruction, merely with much more external energy need than that for the *fission*.

When we say *fusion* corresponds to the *plasma* state it means the sphere symmetrical accelerating elementary collapse.
This starts when all elementary processes have been expanded to the status of the *Hydrogen* process with accumulating proton process intensity potential and electron process *blue shift* impact of infinite low intensity. Without the collapse, the elementary cycle could not be continued.
Plasma is the resolution, the collapse of infinite high intensity– a space-time of infinite high intensity and infinite short process time.

The alternative valid option for finding a new way for energy generation is the acceleration of the *Hydrogen* process in conflict with *gravitation*. As is described in Section 6.

Ref.
S.6

3.6
Important conclusion: the intensities of all proton processes are equal

S.
3.6

With reference to Sections 1, 2 and 3 and to 1G4, 1G5, 2A9 an important conclusion can be made: the intensities of the proton processes of all elementary processes *are quasi equal*.
The intensity of the neutron process has been established by the intensity of the electron process. Neutron processes are "neutral", they are the passive ones.
For calculating the intensity of the electron process the intensities of the proton and neutron processes are taken as measured data, but in fact the intensity of the neutron process depends on the intensity of the electron process.

4
Quantum communication in practice

S.4

Ref.
S.3
S.5

With reference to Sections 3 and 8, elementary processes with **neutron process dominance** mean contribution to *gravitation*. The internal mass/energy intensity surplus potential of the elementary processes is expiring via *gravitation*.

The anti-direction is generating (anti-)electron process *blue shift* surplus. [As physical impact to the quantum membrane, there is no difference: it is about electron process *blue shift* or anti-electron process *blue shift* impact.]

The absolute balance is guaranteed. In intensity terms, the time flow is continuous and the balance is part of the also continuous change/flow/motion/life... processes.

4A1
Ref.
1D4

$$\frac{dmc_x^2}{dt_{a-n}}\left(1-\sqrt{1-\frac{i_x^2}{c_x^2}}\right) \neq \frac{dmc_x^2}{dt_{a-p}}\sqrt{1-\frac{(c_x-i_x)^2}{c_x^2}}\left(1-\sqrt{1-\frac{i_x^2}{c_x^2}}\right); \quad \text{while } E_p = \xi \cdot E_n; \quad e_p \neq e_n$$

4A2

$$\text{with anti-electron process drive } \frac{dmc_x^2}{dt_i\varepsilon_{-xm}}\left(1-\sqrt{1-\frac{(c_x-i_x)^2}{c_x^2}}\right); \text{ with reference to 1D4}$$

The *inflexions* are with the generation of quantum *entropy*. The new elementary process starts with less intensity value.

The anti-processes of the neutron process dominant elementary processes are the drives of the elementary progress,

> ➤ they generate anti-electron process *blue shift* surplus
> ➤ and the surplus "feeds/loads" *gravitation*

resulting in elementary "evolution":

> ➤ the used for *gravitation* anti-electron process *blue shift* surplus is in decreasing tendency and the intensity of the electron process *blue shift* impact of the direct elementary process is also with decreasing gradient: the intensity of the anti-proton collapse corresponds to the intensity of the anti-electron process *blue shift* drive and the intensity of the proton process, in line with this, generates electron process *blue shift* impact of less intensity.

4A3

The intensity of the electron process *blue shift* drive of the neutron process dominant elementary processes is more than the intensity of the anti-electron process *blue shift* drive of the anti-proton process. $\varepsilon_x = \dfrac{1}{\varepsilon_{-mx}}$

(For increasing the intensity - the intensity coefficient shall be of less value.)

In intensity terms, the process goes on by driving the neutrons by increased intensity while the anti-process remains behind with anti-electron process *blue shift* surplus of reduced intensity, as 4A3 proves.

4A4

$$\eta\frac{dmc_x^2}{dt_i\varepsilon_x}\left(1-\sqrt{1-\frac{(c_x-i_x)^2}{c_x^2}}\right); \quad \text{and} \quad \eta\frac{dmc_x^2}{dt_i\varepsilon_{-xm}}\left(1-\sqrt{1-\frac{(c_x-i_x)^2}{c_x^2}}\right)$$

4A5

$$\eta\frac{dmc_x^2}{dt_i\varepsilon_{-xm}}\left(1-\sqrt{1-\frac{(c_x-i_x)^2}{c_x^2}}\right) = \eta\frac{dmc_x^2}{dt_i\frac{1}{\varepsilon_x}}\left(1-\sqrt{1-\frac{(c_x-i_x)^2}{c_x^2}}\right) = \eta\varepsilon_x\frac{dmc_x^2}{dt_i}\left(1-\sqrt{1-\frac{(c_x-i_x)^2}{c_x^2}}\right);$$

4A4 and 4A5 mean:

there are here η number of neutron processes (driven by ε_x intensity) and η number of anti-proton processes, driven by $\varepsilon_{-xm} = 1/\varepsilon_x$ intensity, and $\varepsilon_x < 1$.

$\varepsilon_{-xm} > \varepsilon_x$ so, the intensity of the anti-electron process is less than the intensity of the electron process. The intensity of the anti-neutron process is equal to the intensity of the neutron process. The η number of the electron and the neutron processes is the consequence of the generating η number of proton processes within the direct cycle without electron process surplus.

In line with 4A4 and 4A5 it follows:

- The generation of the electron and the anti-electron processes are equal to η - but these equal numbers are generating for different time periods. Meaning: they are of different intensities.
- The generation of η number of anti-electron processes takes less time, as the neutron and anti-neutron processes are of more intensity than the intensities of the anti-proton and the proton processes.
- The intensity of the utilisation of the electron process is equal to

  $\dfrac{\eta}{\varepsilon_x}$ - as for the neutron process dominant elementary processes the coefficient is $\varepsilon_x < 1$ - it means high intensity use, reference to 4A4;   4A6

  $\varepsilon_x \eta$ - as for the anti-proton process the utilisation is of less value, reference to 4A5; the anti-electron process generation is η; the utilisation is with clear   4A7

  surplus of $\Delta = \dfrac{\eta}{\varepsilon_x} - \mu\varepsilon_x = \eta(\dfrac{1}{\varepsilon_x} - \varepsilon_x)$ for loading *gravitation*.   4A8

Neutron process dominant elementary processes leaving the *Earth* are ready for elementary communication, establishing balanced or quasi balanced elementary processes.

Elementary processes with electron process *blue shift* surplus and **proton process dominance** operate in the same way, just that the electron process *blue shift* surplus is generating on the other, direct side.

$$\frac{dmc_x^2}{dt_p}\left(1 - \sqrt{1 - \frac{i_x^2}{c_x^2}}\right) \neq \frac{dmc_x^2}{dt_n}\sqrt{1 - \frac{(c_x - i_x)^2}{c_x^2}}\left(1 - \sqrt{1 - \frac{i_x^2}{c_x^2}}\right); \text{ while } E_{a-p} = \xi \cdot E_{a-n}; \quad e_{a-p} \neq e_{a-n} \quad \text{4B1}$$

the drive of the neutron process is: $\dfrac{dmc_x^2}{dt_i \varepsilon_x}\left(1 - \sqrt{1 - \dfrac{(c_x - i_x)^2}{c_x^2}}\right)$ 4B2

The electron process *blue shift* surplus has direct impact on the quantum communication. Proton process dominant elementary processes use their electron process *blue shift* surplus for initiating elementary communications.

The **proton** and the **neutron process dominance** raises the question, how can an electron process, product of the proton process, drive a neutron process, the intensity of which is higher than the intensity of the proton process?

The answer is coming from the integrated character of the direct and the anti-processes.

In absolute terms the balance is equally guaranteed, as for the direct and the anti-processes:

$$\frac{dmc^2}{dt_p \varepsilon_p}\left(1 - \sqrt{1 - \frac{v^2}{c^2}}\right) = \xi \frac{dmc^2}{dt_p \varepsilon_p}\sqrt{1 - \frac{(c - i)^2}{c^2}}\left(1 - \sqrt{1 - \frac{v^2}{c^2}}\right) \quad \text{4B3}$$

In intensity terms, the quantum membrane of the elementary process is one and the same equally for the electron and also for the anti-electron processes: the quantum speed is the same for both, process and anti-process lines:

The intensity of the quantum membrane, established by the drives, is:

4B4
$$e = \frac{dmc^2}{dt_i \varepsilon}\left(1 - \sqrt{1 - \frac{(c-i)^2}{c^2}}\right) + \frac{dmc^2}{dt_i \frac{1}{\varepsilon}}\left(1 - \sqrt{1 - \frac{(c-i)^2}{c^2}}\right) = \left(\frac{1}{\varepsilon} + \frac{\varepsilon}{1}\right)\frac{dmc^2}{dt_i}\left(1 - \sqrt{1 - \frac{(c-i)^2}{c^2}}\right)$$

4B5

The integrated acting intensity ε_{ex} of the quantum membrane of the element is:	$\varepsilon_{ex} = \frac{\varepsilon}{\varepsilon^2 + 1}$; as $\varepsilon_e = \frac{1+\varepsilon^2}{\varepsilon}$;	$\varepsilon_{ex} < 1$ in any circumstances

If one includes the value of the integrated intensity in 4B5 into the intensity formula with the acting intensity, the integrated electron process gives:

4B6
$$e = \varepsilon_e \frac{dmc^2}{dt_i}\left(1 - \sqrt{1 - \frac{(c-i)^2}{c^2}}\right) = \frac{dmc^2}{dt_i \varepsilon_{ex}}\left(1 - \sqrt{1 - \frac{(c-i)^2}{c^2}}\right); \quad \text{and} \quad \varepsilon_{ex} = \frac{1}{\varepsilon_e} < \frac{1}{\varepsilon}$$

The integrated intensity of the quantum membrane, generated by the electron processes, including the direct and also the anti-process directions, is higher than the intensity of the directions separately. (The de facto intensity value is inversely proportional to the value of ε, the intensity coefficient.)

In this way, the intensities of the direct and the (anti-)electron process *blue shift* impacts drive the neutron and the anti-proton processes to full collapse in any circumstances. The surplus either feeds *gravitation* or initiates elementary communication.

The exemptions are the *Hydrogen* process, where $\varepsilon = \infty$ and ε_{ex} in 4B5 and 4B6 is with $\lim \varepsilon_{eH} = \infty$ gives indeterminate result and the *plasma* state where the integrated intensity with $\lim \varepsilon_{pl} = 0$ is approaching $\lim \varepsilon_{epl} = 0$ infinite value.

S.
4.1

4.1.
Mass/energy intensities to be utilised

Ref
4A8

4C1

The controlling function of the anti-neutron process means that while the neutron process is the passive and the driven one, the anti-neutron process with reference to 4A8 is controlling the existence of the elementary related Δ_x anti-electron process surplus. It will either be restored or modified by new elementary Δ_y surplus.

The other specific characteristic of the neutron process intensity dominant elementary processes is that the anti-proton process remains "behind": While the anti-electron process is with surplus, its intensity is of reduced value.

The communication of neutron process dominant elementary processes with proton process dominant elementary processes is easy as the processes of the two communicating sides are symmetrical and supplementing each other.

In proton process dominant elementary processes, the anti-electron process *blue shift* impact is the one of increased intensity and the surplus is accumulating on the direct side.

The elementary evolution is approaching a mass/energy intensity status when the intensity capacity of the proton process and the generating intensity of the electron process *blue shift* impact is only capable of driving a collapse of equal, or close to equal, or reduced – relative to the proton expansion – intensity.

$$\frac{dmc_x^2}{dt_i\varepsilon_x}\left(1-\sqrt{1-\frac{(c_x-i_x)^2}{c_x^2}}\right)$$ and from the proton/neutron process relation $\varepsilon_x = \frac{\varepsilon_{xp}}{\varepsilon_{xn}}\sqrt{1-\frac{(c_x-i_x)^2}{c_x^2}} > 1; \quad \frac{dt_{xn}}{dt_{xp}} > 1$ 4C2

The time consumption of the driven neutron process is clearly representing the capacity of the proton process.

At the same time, the anti-neutron and anti-electron processes are restoring the elementary process. It is obligatory, since otherwise the elementary relation would be over.

There is no surplus in the anti-electron process *blue shift* impact within a proton process dominant case. It is of increased intensity: $\varepsilon_{-xm} = \frac{1}{\varepsilon_x} < 1$ 4C3

There are two types of proton process dominant elementary processes:
Ref. Table 1.1
  ➢ elementary processes with original speed value of quantum communication (reference to Table 1.1) equal to or more than the speed value on the *Earth* surface: *Carbon, Silicon, Sulphur, Calcium;* and
  ➢ the ones with their original speed of quantum communication less than the value on the *Earth* surface 299,792 km/sec: *Oxygen, Nitrogen, Helium* and *Hydrogen.* Being part of the *Earth' quantum system of reference* these elementary processes communicate with the quantum speed of the *Earth.* This makes these elementary processes conflicting, gaseous and liquid.

The benefit of both the proton process dominant and the neutron process dominant elementary processes is that they establish the conditions of their intensive quantum communication.

Proton process dominancy is with weak electron process *blue shift* impact but with surplus, neutron process dominancy is with electron process *blue shift* impact of increased intensity and with surplus on its anti-elementary side.
Ref. S.3
The elementary mass/energy transfer with reference to Section 3 needs three cycles.

The formulating electron processes cannot drive themselves. As the process continues, electron processes drive the neutrons of the following cycle.

The mass/energy transfer within the elementary processes, covered by the proton process and the anti-neutron processes, happens within the same cycle.

4.2.
S. 4.2
The way of the formulation of the surplus

$$\varepsilon_e = \frac{\varepsilon_p}{\varepsilon_n}\sqrt{1-\frac{(c_x-i_x)^2}{c_x^2}} = \frac{dt_n}{dt_p}; \text{ meaning } \frac{dt_n}{dt_p} > 1 \text{ or } \frac{dt_n}{dt_p} < 1$$ This decides the time relation (duration) of the processes.   4D1

The intensities of the proton or neutron processes decide the elementary relation.

The proton/neutron intensity relations are indeed about relations. In the majority of the cases, the intensity of the proton process of the proton process dominant elementary processes is less than the intensity of the proton process of the neutron process intensive elementary processes. The intensity increase of the proton process is, in fact, not about real intensity increase rather about the relation: the closer to the *plasma* the anti-neutron process (and the neutron process) is, the more intensive it is, as the internal energy is more.

  The best example is the *Hydrogen,* which is not about infinite high proton process intensity, rather, about infinite low neutron process (and anti-neutron process) intensity.

S.
4.3

4.3.
Quantum communication of elementary processes

Quantum communication of the elementary process "A" and the elementary process "B" means the resulting process is closer to the elementary balanced state than the two elementary processes originally and separately are.

> the electron process *blue shift* surplus of elementary process A is used for driving the neutron process of B – otherwise driven by B by high intensity, while

> the anti-electron process *blue shift* surplus of elementary process B is used for driving the anti-proton process of A – otherwise driven by A by higher intensity.

Neutron and anti-proton processes are of passive character.

A: B:

4E1	$\dfrac{dmc_A^2}{dt_i \varepsilon_A}\left(1-\sqrt{1-\dfrac{(c_A-i_A)^2}{c_A^2}}\right)$	$\dfrac{dmc_B^2}{dt_i \varepsilon_B}\left(1-\sqrt{1-\dfrac{(c_B-i_B)^2}{c_B^2}}\right)$	$\varepsilon_B < \varepsilon_A;\quad \varepsilon_A > 1;\ \varepsilon_B < 1;$ $\varepsilon_A = n\cdot\varepsilon_B;\ \varepsilon_B = \varepsilon/n;$ It is taken that: $\varepsilon_A = \varepsilon$
4E2	$\dfrac{dmc_A^2}{dt_i \varepsilon_{a-A}}\left(1-\sqrt{1-\dfrac{(c_A-i_A)^2}{c_A^2}}\right)$	$\dfrac{dmc_B^2}{dt_i \varepsilon_{a-B}}\left(1-\sqrt{1-\dfrac{(c_B-i_B)^2}{c_B^2}}\right)$	$\varepsilon_{a-B} > \varepsilon_{a-A};\ \varepsilon_{a-A} < 1;$ $\varepsilon_{a-B} > 1;\ \varepsilon_{a-B} = n\cdot\varepsilon_{a-A};$ $\varepsilon_{a-A} = 1/\varepsilon;\quad \varepsilon_{a-B} = n/\varepsilon$

The closer the proton/neutron and anti-neutron/anti-proton relations are to each other, the more efficient is the result.

If the proton/neutron process intensity relation is increased, more volume of elementary process A is necessary for having the best result.

The goal is to have the resulting intensities of the proton and the neutron processes and also the anti-neutron and the anti-proton processes closer to each other. The absolute balance would result in unbreakable structure.

Ref
S.1
S.2

With reference to Section 1 and 2 and the rule of $c^2 \cdot \varepsilon = const$ the consequence of the difference in the speed values of quantum communication modifies the intensity relations of elementary processes.

4E3 With reference to 4E1 and 4E2, if it is taken that $c_B > c_A$ and $x = \dfrac{c_B^2}{c_A^2}$

With the view that it is taken $\varepsilon_A = \varepsilon$, the equations could be written as:

4E4 $\dfrac{dmc_A^2}{dt_i \varepsilon}\left(1-\sqrt{1-\dfrac{(c_A-i_A)^2}{c_A^2}}\right)$; and $\dfrac{dmc_B^2}{dt_i \frac{\varepsilon}{n}}\left(1-\sqrt{1-\dfrac{(c_B-i_B)^2}{c_B^2}}\right) = x\cdot n\dfrac{dmc_A^2}{dt_i \varepsilon}\left(1-\sqrt{1-\dfrac{(c_A-i_A)^2}{c_A^2}}\right)$

and

4E5 $\dfrac{dmc_A^2}{dt_i \frac{1}{\varepsilon}}\left(1-\sqrt{1-\dfrac{(c_A-i_A)^2}{c_A^2}}\right)$ and $\dfrac{dmc_B^2}{dt_i \frac{n}{\varepsilon}}\left(1-\sqrt{1-\dfrac{(c_B-i_B)^2}{c_B^2}}\right) = \dfrac{x}{n}\dfrac{dmc_A^2}{dt_i \frac{1}{\varepsilon}}\left(1-\sqrt{1-\dfrac{(c_A-i_A)^2}{c_A^2}}\right)$

4E4 shows the direct relation:

for A to be in electron process *blue shift* impact intensity balance with B its volume must

4E6 be $n\cdot x$-times more: $B = x\cdot n\cdot A$

4E5 proves the anti-process:

B works in anti-electron process *blue shift* intensity balance with A if the volume of A is

(x/n)-times more: $a\text{-}B = \dfrac{x}{n} a\text{-}A$ 4E7

The quantum speed value is the same and equal for the direct and anti-processes. 4A7 is also correct, since the intensity of the anti-electron process of A is equal to $\quad \varepsilon_{a\text{-}A} = \dfrac{1}{\varepsilon_{a\text{-}A}} = \dfrac{1}{\varepsilon}$ 4E8

So, the direct way gives: B is $(x \cdot n)$-times more than A

the anti-direct way gives B is (x/n)-times more than A.

The only valid solution is $n = 1$ and, consequently, $x = 1$ as well, meaning A and B represent the same elementary processes.   4E9

It seems there is no way for elementary communication.

But if it is taken that:

X is the intensity of the generation of the electron process of A and

Y is the intensity of the generation of the electron process of the elementary process of B.

It is also taken that

- intensity portion $(X - A)$ of the electron process *blue shift* impact of A is working internally within for keeping the internal balance of A and
- intensity portion $(Y - B)$ works internally for the balance of B.

A is about the intensity of the electron process establishing the balance with B on direct way. So, $(X - A) + A = X$ 4F1

B is the intensity of the anti-electron process establishing balance with A on the anti-direction. So, $(Y - B) + B = Y$ 4F2

Electron processes working on the internal balance mean $n = 1$ and $x = 1$ similarly for A and B.

With reference to the conditions above and

to 4E4 corresponding to the direct relation:	to 4E5 corresponding to the anti-direction:
$(X - A) + A = 1 + n \cdot x = X :$	$(Y - B) + B = 1 + \dfrac{n}{x} = Y ;$

     4F3
     4F4

[With reference to 4E9 – elementary processes themselves are in full balance (this is the reason for *1* in 4F3 and 4F4) – plus the communicating portions give the whole intensity of the electron process generation.]

x can be calculated in line with 4E3 as the quantum speed data of the elementary processes to be mixed are known, reference to Table 1.1.

n also can be calculated in line with 4E1 and 4E2:

$$n = \dfrac{\varepsilon_A}{\varepsilon_B}; \quad \text{as } \varepsilon_A = \dfrac{1}{\varepsilon_{a\text{-}A}} \text{ and } \varepsilon_B = \dfrac{1}{\varepsilon_{a\text{-}B}}; \quad n = \dfrac{\varepsilon_{a\text{-}B}}{\varepsilon_{a\text{-}A}} \text{ indeed.}$$ 4F5

The main points of the deduction and the explanation are that

- the mix of elementary processes shall ensure the internal and the external balance in line with the direct and the anti-processes as well, in line with 4F5; 4E4 and 4E5 are only about the internal balance.
- the relation of the mass/energy exchange between processes and anti-processes of any elementary process is equal to 1 as the exchange even in any not balanced case is equal within the elementary process itself.
- the electron process and the anti-electron process *blue shift* intensity impacts communicate the same mass/energy intensity, as it is presented on Diag.4.2.

In external elementary quantum communication, the relation of the intensities of the direct and the indirect drives makes the difference.

The result gives intensity relation, which is proportional to volumes.

In Exp.4.1 the quantum speed values have also been consolidated since, in line with 4E3, the quantum speed values become equal.

Exp. 4.1

Extract from Table 1.1	P N	Summarised intensity (Measured atomic weight)	ε	c speed of quantum communication	Example
Carbon	6	12.01100	1.01338	*300000*	Mixing *Chlorine* with *Silicon*. Results of the calculation The balanced mix is:
Nitrogen	7	14.00670	1.01420	298878*	$n = \dfrac{0.93467}{1.00898} = 0.926351$
Oxygen	8	15.99900	1.01533	299711*	
Fluorine	9	18.99840	0.91316	316033	$x = \dfrac{312376^2}{300652^2} = 1.03895$
Neon	10	20.17000	0.99810	302286	$\dfrac{n}{x} = 0.891584$
Sodium (Na)	11	22.98900	0.93084	313018	
Magnesium	12	24.30500	0.98990	303536	$n \cdot x = 0.962475$
Aluminium	13	26.98150	0.94697	310340	
Silicon	14	28.08550	1.00898	*300652*	$Y = 1 + (n/x) = 1.891584$
Phosphorus	15	30.97370	0.95283	309348	$X = 1 + n \cdot x = 1.962475$
Sulphur	16	32.06000	1.01143	*300288*	The relation of the mix as per
Chlorine	17	35.45300	0.93467	312376	A/B = *Silicon/Chlorine*
Argon	18	39.94800	0.83139	331210	should be: $\dfrac{Y}{X} = \underline{1.037477}$
Potassium (K)	19	39.09800	0.95932	308335	
Calcium	20	40.08000	1.01112	*300334*	Exp.4.1

Elementary processes with electron process *blue shift* surplus are: *Hydrogen, Helium, Carbon, Oxygen, Nitrogen, Silicon, Sulphur* and *Calcium* only.

Nature provides us with elementary processes, with electron and anti-electron *blue shift* impacts and surplus, we just need to use them for the benefit of all of us.

S. 4.4

4.4.
Anti-processes

The direct and anti-processes happen in parallel:

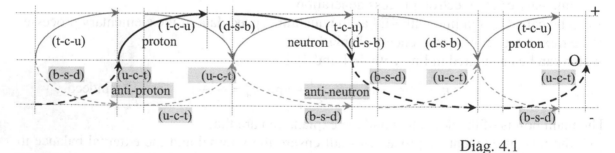

Diag. 4.1

Diag. 4.1

Processes have their anti-direction controlling component.

There is no proton process without anti-proton process and no neutron process without anti-neutron process.

There is no (t-c-u) quarks process without (u-c-t) and (d-s-b) without (b-s-d).

Anti-processes are establishing the "background" of the elementary status:

➤ (t-c-u) expansion happens by (u-c-t) preparatory stage;

➤ (d-s-b) collapse runs out by (b-s-d) chain.

Anti-processes are the tools of the fluent transition from collapse to expansion.

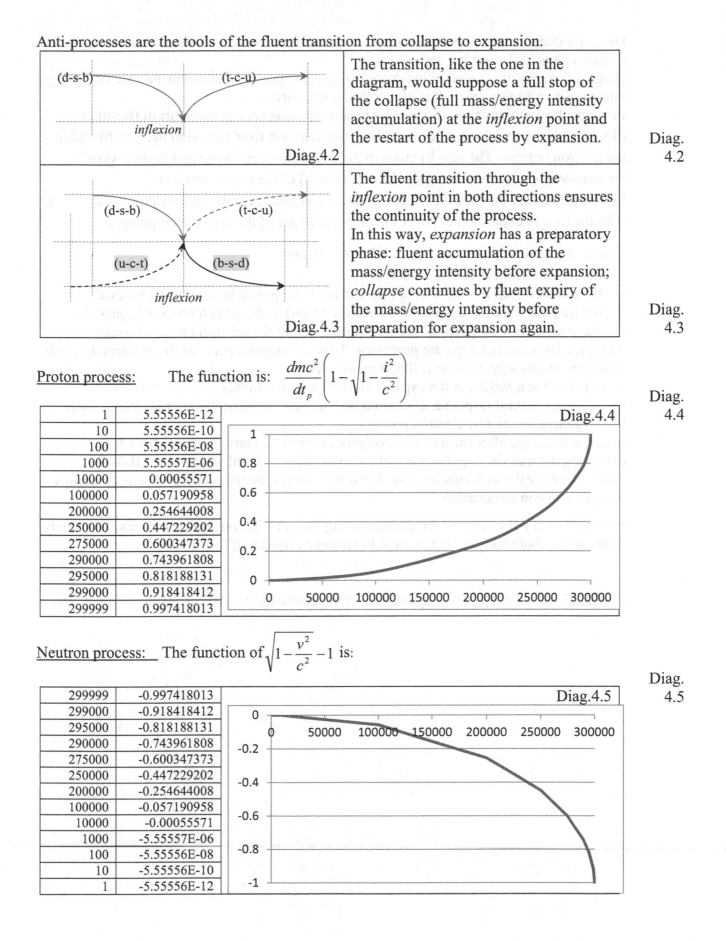

| | The transition, like the one in the diagram, would suppose a full stop of the collapse (full mass/energy intensity accumulation) at the *inflexion* point and the restart of the process by expansion. | Diag. 4.2 |

Diag.4.2

| | The fluent transition through the *inflexion* point in both directions ensures the continuity of the process. In this way, *expansion* has a preparatory phase: fluent accumulation of the mass/energy intensity before expansion; *collapse* continues by fluent expiry of the mass/energy intensity before preparation for expansion again. | Diag. 4.3 |

Diag.4.3

Proton process: The function is: $\dfrac{dmc^2}{dt_p}\left(1-\sqrt{1-\dfrac{i^2}{c^2}}\right)$

Diag. 4.4

1	5.55556E-12
10	5.55556E-10
100	5.55556E-08
1000	5.55557E-06
10000	0.00055571
100000	0.057190958
200000	0.254644008
250000	0.447229202
275000	0.600347373
290000	0.743961808
295000	0.818188131
299000	0.918418412
299999	0.997418013

Diag.4.4

Neutron process: The function of $\sqrt{1-\dfrac{v^2}{c^2}}-1$ is:

Diag. 4.5

299999	-0.997418013
299000	-0.918418412
295000	-0.818188131
290000	-0.743961808
275000	-0.600347373
250000	-0.447229202
200000	-0.254644008
100000	-0.057190958
10000	-0.00055571
1000	-5.55557E-06
100	-5.55556E-08
10	-5.55556E-10
1	-5.55556E-12

Diag.4.5

Diag. 4.4 demonstrates the accelerating expansion for the count of the internal energy/mass intensity. It is of decreasing intensity.

Diag. 4.5 shows the accelerating mass/energy collapse for the count of external energy intensity (electron process). It is of increasing intensity.

* In elementary weights*, proton process means the mass/energy transport of (t-c-u) to (d-s-b) within the proton process and the same transport from (t-c-u) to the (d-s-b) within the neutron process. The (d-s-b) chain in the neutron process is the one to be covered by the mass/energy intensity transport from the (t-c-u) of the proton process.

In anti-processes, the calculation is the same, just that the anti-neutron process is the one with the (b-s-d) expansion to cover the (u-c-t) collapse of the anti-proton process.

* Proton and neutron process are with 3-3 unit weights*:
- the transfer from (t-c-u) to (d-s-b) is 1;
- the transfer from (b-s-d) to (u-c-t) is 1 in both, the proton and neutron processes;
- the transfer from the neutron (b-s-d) to the (u-c-t) of the proton process is quasi 1;
- the transfer from the proton (t-c-u) to the (d-s-b) of the neutron process is quasi 1.

Anti-processes are not separate processes. They are organic parts of the elementary cycle, fulfilling, in this way, the controlling function also.

(u-c-t) after the *inflexion* is the expansion of the collapsed mass/energy intensity.

(b-s-d) is the natural response, expansion and the generation of the internal energy intensity by the generation of anti-electron process.

(u-c-t) is the stage after the anti-electron process, which needs *blue shift* drive for collecting the mass/energy intensity of the expansion. Full utilisation of the (b-s-d) expansion after the *inflexion* by *blue shift* drive ensures the mass/energy intensity surplus for quantum communication.

* * weight, as a definition, is for demonstrating the case in conventional terms. Obviously, this is not about weight, rather about the intensity impact of the process.

Explanations and examples of quantum communication in practice is given in Annex 4.1.

5
Space-time matrix

The *inflexion* is missing in the *Hydrogen* process: the electron process is driving and the proton process is covering the neutron process of infinite low intensity – without full collapse. The *plasma* is the full collapse with *inflexion* of the neutron process, continuing by anti-neutron process, both of infinite high intensities.

The cycle of the elementary evolution is reaching its completion stage when all elementary processes have been turned into *Hydrogen* process. The neutron processes – under the impact of the still acting electron process *blue shift* drives of infinite low intensity – are collapsing in infinite numbers.

The space-time is full of *Hydrogen* processes, all having electron process *blue shift* impact of infinite low quantum speed. There are three elementary lines running in parallel: the neutron process of infinite low intensity; the electron process *blue shift* drive of infinite low intensity and the proton processes with infinite high quantum cover potential, because of the non-use. This is presented below in Diag.5.1.

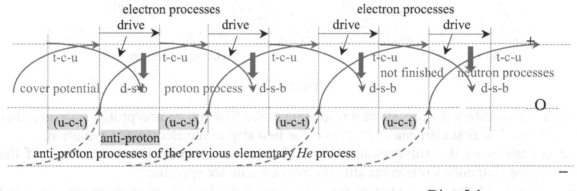

Diag. 5.1

This is the reason that the cover potential of the proton process of the *Hydrogen* process at this final stage is of infinite high intensity. This potential has been accumulating for the whole period of the elementary evolution without its balancing use. At this completion stage it finally initiates the neutron collapse. This is not just about the intensity difference between the proton and the neutron processes, but also about the absolute balance of the elementary evolution.

$$\frac{dmc^2}{dt_p \varepsilon_p}\left(1-\sqrt{1-\frac{v^2}{c^2}}\right) = \xi \frac{dmc^2}{dt_n \varepsilon_n}\sqrt{1-\frac{(c-i)^2}{c^2}}\left(1-\sqrt{1-\frac{v^2}{c^2}}\right)$$

5A1

this part was completely missing

The collapse is irrevocable and goes with infinite high intensity, until the *inflexion* stops it. This *inflexion* is the *plasma* status itself, with neutron and anti-neutron processes of infinite high intensities. The key points here are:
-   the infinite high quantum speed of the drive of the collapse; and as its consequence
-   the infinite high intensities of the neutron and the anti-neutron processes.

The quantum communication here is about the quantum connection between the proton and the neutron processes; between the energy source potential and the energy demand. The proton process potential is acting through the electron process. The electron process drives the neutron process, resulting in infinite high quantum speed and intensity.

The anti-elementary side of the elementary *Hydrogen-plasma* communication starts with the anti-neutron process of the *plasma*. *This also means the start of the elementary evolution.*

The anti-neutron process of the *plasma* is generating anti-electron process *blue shift* impact of infinite high intensity. The intensity of the drive of the anti-proton collapse, however, shall correspond to the intensity need of the proton process of the *first* elementary process with close to zero intensity. Therefore, the anti-electron process of the *plasma* is in infinite high surplus.

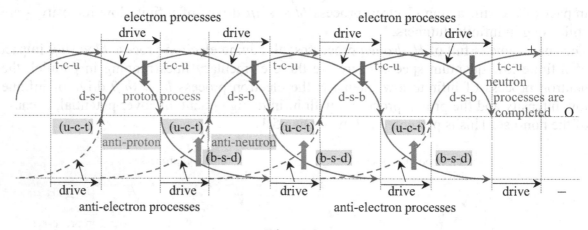

Diag.
5.2

Diag.5.2

The first elementary process starts formulating at the first anti-proton/proton *inflexion* after the *plasma*. The first elementary process is the first step of the elementary evolution.

The intensities of the anti-proton/proton and the neutron/anti-neutron *inflexions* of the formulating elementary processes after the *plasma* state are specific:

The intensity of the anti-proton/proton *inflexion* of the first elementary process is quasi *zero*. The gradient of the intensities of the *inflexions* of the following elementary processes is of increasing tendency. The intensity of the neutron/anti-neutron *inflexion* of the first elementary process is quasi equal to the infinite high intensity of the *plasma* status. The gradient of the intensities of the *inflexions* of the following elementary processes is of decreasing tendency.

Approaching the *Hydrogen* process, the intensity of the neutron/anti-neutron *inflexion* is of infinite low intensity.

Ref.
S.1
S.2

The intensity of the anti-proton process defines the intensity of the proton process on the direct side. The evolution continues, in line with the rule of the $c_x \cdot \varepsilon_x = const$

(Reference to Section 1 and 2.)

The surplus of the anti-electron process *blue shift* impact of the *plasma* and of the elementary processes of the *post-plasma* state are of infinite high value and conflict, generating infinite high temperature. With reference to Diag.5.3, the elementary formations after the *plasma* are in fact parts of the *plasma* state. Because of the acting infinite high anti-electron process *blue shift* conflict, the *plasma* and the after *plasma* stages are of infinite high energy/intensity potential and temperature.

All this generating infinite high intensity potential of the anti-neutron processes shall be processed by the elementary processes of the elementary evolution.

As the evolution progresses, the conflict becomes less and less. The first elementary processes of the *Earth* core/mantle are around intensity value of $\varepsilon_x \approx 0.6$ and more. For example, for the *Neptunium* process, it is $\varepsilon_{Np} = 0.65359$.

The anti-electron process *blue shift* surplus of the elementary processes of the close to the *post-plasma* state is the main source of loading/feeding *gravitation*. This *post-plasma* state means the scale of the intensity coefficient: $\propto < \varepsilon_x < 0.5$.

Ref.
Table
1.1

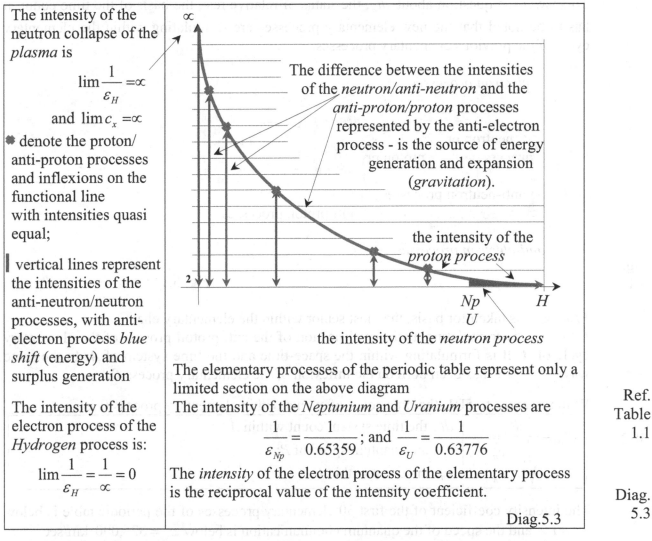

The intensity of the neutron collapse of the *plasma* is

$$\lim \frac{1}{\varepsilon_H} = \propto$$

and $\lim c_x = \propto$

✳ denote the proton/anti-proton processes and inflexions on the functional line with intensities quasi equal;

❘ vertical lines represent the intensities of the anti-neutron/neutron processes, with anti-electron process *blue shift* (energy) and surplus generation

The intensity of the electron process of the *Hydrogen* process is:

$$\lim \frac{1}{\varepsilon_H} = \frac{1}{\propto} = 0$$

The difference between the intensities of the *neutron/anti-neutron* and the *anti-proton/proton* processes represented by the anti-electron process - is the source of energy generation and expansion (*gravitation*).

the intensity of the *proton process*

the intensity of the *neutron process*

The elementary processes of the periodic table represent only a limited section on the above diagram

The intensity of the *Neptunium* and *Uranium* processes are

$$\frac{1}{\varepsilon_{Np}} = \frac{1}{0.65359}; \text{ and } \frac{1}{\varepsilon_U} = \frac{1}{0.63776}$$

The *intensity* of the electron process of the elementary process is the reciprocal value of the intensity coefficient.

Diag.5.3

Ref.
Table
1.1

Diag.
5.3

The intensities of the anti-proton/proton *inflexions* of the elementary processes are equal, since the anti-electron process IQ_- drive for all elementary processes is equal.

The cycles of the elementary evolution in Diag.5.3 are restarting and continue with the constant generation of quantum impulses (quantum) – until the entire elementary potential turns into quantum impulse. As with reference to Sections 1 and 2: $c_x \cdot \varepsilon_x = const$

Ref.
S.1

This relation can be constant if $c_x = k \cdot \dfrac{1}{\varepsilon_x}$;

5A2

Meaning: The value of the quantum speed is proportional to the intensity (to the inverse of the intensity coefficient) of the elementary process.

S.
5.1

Ref.
3A4
5B1

5.1
Elementary space-times are embedded within each other
and formulating *matrix*

$$dt_x = \frac{dt_p = dt_o = dt_n}{\sqrt{1 - \dfrac{v_x^2}{c_x^2}}}$$

c_x with reference to 5A2 above is the speed value of the quantum communication, function of the elementary process;

v_x is the event;

dt_o the status of relative rest was always with question mark.

To answer the question about dt_o, the status of relative rest, the basis of the time count, it has to be noted that the new elementary processes are formulating within the elementary cycle of the previous elementary processes.

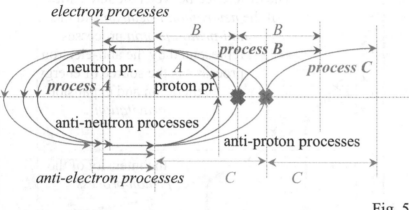

electron processes

neutron pr.

process A   proton pr

anti-neutron processes

anti-electron processes

process B

process C

anti-proton processes

Fig.
5.1

Fig. 5.1

Process A is taken for basis, the most senior within the elementary chain

Process B is formulating as the modification of the anti-proton process of the elementary cycle of *A*. It is formulating within the space-time and the time system of *A*. In the same way, *process C* has been born on the inflexion of the elementary process **B**.

With reference to 5B1, the time count of events within elementary process B is:

5B2

$dt_B = \dfrac{dt_A}{\sqrt{1 - \dfrac{v_B^2}{c_B^2}}}$	dt_A the time system/count within A; c_B the quantum speed of B; v_B the event within B.

The intensity coefficient of the first 30 elementary processes of the periodic table is below $\varepsilon_x = 1.2$ and the speed of the quantum communication is below $c_x = 326,000$ km/sec.

Thus, there are no significant differences in space-times and time counts even in the case of high speed elementary events. Differences may occur if

- considering elementary processes with increased quantum speed and electron process intensity, corresponding to the second half of the periodic table, or
- the quantum speed is increased, as result of electron process *blue shift* conflict.

Elementary processes establish a certain space-time matrix having built up from *plasma* with the highest intensity to the *Hydrogen* process, the one with the lowest intensity.

5.2
Communication of space-times

Elementary processes are in quantum communication as
   (1) the time systems of all electron process drives are quasi equal;
   (2) all neutron processes are passive and driven; and in addition
   (3) the quantum drive of all elementary anti-electron processes is equal.

These above conditions also mean that the intensities of the proton processes for all elementary processes are equal. The intensity of the drive of the neutron process is the one which determines the intensity relation and the type of the elementary process.

Elementary processes generate their own space-time as the speed values of their quantum communication are different.
The space-time matrix contains the space-time of all elementary processes built up on each other.
The *space* contains all quantum impulses taken altogether as an aggregate system. In other words, *space* is about the system of all quantum impulses, (quantum – event of infinite low intensity) transporting quantum impacts. Elementary processes impact the quantum system (space) in each direction. Quantum impulses transfer quantum signals in any direction.
The transportation of quantum impacts and signals is the quantum communication itself.
Quantum impulses (the energy quantum) do not add anything to the impacting and the transported quantum signals.
As the *space-time* is about quantum impulses (energy quantum), the key indicator of its "size" is the value of the quantum speed.

    The quantum speed determines its acting intensity and establishes the acting quantum drive within the space-time: $IQ_x = \dfrac{c_x^2}{\varepsilon_x}$

The space-time of elementary processes is characterised by the quantum speed; all other parameters are consequences: $c_x^2 \cdot \varepsilon_x = \dfrac{c_x^2}{\varepsilon_{xm-}} = const$

The distance, as such, for the measurement of space has "only" practical importance.
The definition of the space by length dimensions is the conventional way of measuring it.
Space-time is about quantum impacts with no limitation of intensity and time!
Higher quantum speed means higher intensity, approaching more quantum.

Thus, the definition of *space* is given by events!
The same way as it is for the definition of *time*, merely that time measurement is about a single *event*; space measurement means an infinite number of *events*.

    Quantum signals are electron process *blue shift* impacts with specific quantum speed and intensity, having the quasi common time system of the impact: $e_x = \dfrac{dmc_x^2}{dt_i \varepsilon_x}\left(1 - \sqrt{1 - \dfrac{(c_x - i_x)^2}{c_x^2}}\right)$

The transfer by the quantum system of this impact is actually the *space-time* itself.
If this impact is a separate signal, it makes no sense to call it as space-time. Space-time means acting in parallel similar quantum signals of equal quantum speed. With reference to 1N5, the value of the quantum speed of the impact is the distinguishing criteria.

The quantum speed is the speed of the elementary communication. Each elementary process has its own quantum speed. Each elementary process has its own space-time.
Space-times, like the space-time of the *Earth* surface, contain infinite number of quantum signals with the variety of intensities, but all of them having equal quantum speed value.

Quantum impulses equally transfer quantum impacts with different quantum speed values. The explanation is the following:

➤ All quantum impacts have the same dt_i time system.

➤ The quantum impulse is electron process *blue shift* impact of infinite low intensity:
  - any external signal, therefore, is treated by conflict;
  - quantum does not take impacts and remain of infinite low intensity;
  - quantum does transfer signals promptly without modification and time delay:

➤ The intensity of the impacting signal is the one which is generating the intensity of the conflict.

➤ Quantum impulses (quantum) equally transfer all impacts with no difference in their acting quantum speed.

➤ Quantum signals with different quantum speed values impact *a different number* of quantum impulses (for dt_i the unit period of time, which is quasi equal for all quantum signals)
  - space-time with quantum speed c_x is in quantum communication, impacting n_x quantum;
  - all signals of the quantum communication within this space-time of this quantum speed, whatever the events are about, always impact n_x quantum;
  - at the same time, the intensities of the events may vary on infinite scale;
  - the intensity of the event is the one which determines the intensity of the conflict by each of the equally impacted number of quantum:

5C4
$$\frac{dmc_x^2}{dt_i \varepsilon_x} + n_x \frac{dmc_q^2}{dt_i \varepsilon_q} = \frac{dmc_x^2}{dt_i \varepsilon_x}; \qquad \text{as} \qquad \lim \frac{c_q^2}{\varepsilon_q} = 0$$

ε_x is the one which is variant within the space-time of certain quantum speed.

➤ The *space-time* establishes its own time system:
  - impacting certain number of quantum means *intensity*;
  - the intensity directly relates and identifies *time*;
  - there is a difference, with reference to 5C4, between ε_x the intensity of the impact and n_x the number of the impacted quantum within dt_i the common time system of all electron process blue shift impacts:
  - the time system of the space-time means the number of the impacted quantum for the unit period of time:

5C5
$$dt_x = \frac{dt_i}{n_x}; \quad \text{and} \quad n_x = \frac{dt_i}{dt_x}$$

  - meaning: the *space* (= the quantum system) is the same, with the distinguishing parameter of the *time* (= the number of the impacted quantum);

➤ *Space-times* exist in parallel:
  - different space-times means having quantum impacts in the quantum system by the variety of quantum speed values:

5C6
$$dt_{x1} = \frac{dt_i}{n_{x1}}; \quad dt_{x2} = \frac{dt_i}{n_{x2}}; \dots dt_{xn} = \frac{dt_i}{n_{xn}}$$

  - higher speed of quantum communication means space-time of increased intensity, shorter in time flow;

➤ In a *space-time* of increased intensity (= with increased quantum speed value = with increased number of impacted quantum) the time flows *more slowly*:

Ref.
5C5
5C7

$$\text{with reference to 5C5} \qquad f(c_x) = f(n_x) = f\left(\frac{1}{\Delta t_x}\right)$$

➤ One and the same quantum, impacted by quantum signals of different quantum speed, does not have any conflicting effect on the quantum. It stays of infinite low intensity value as it was. The quantum signals are the ones which will be in conflict.
With reference to 5C4 earlier,

$$\frac{dmc_{x1}^2}{dt_i \varepsilon_{x1}} + n_{x1}\frac{dmc_q^2}{dt_i \varepsilon_q} + \frac{dmc_{x2}^2}{dt_i \varepsilon_{x2}} + n_{x2}\frac{dmc_q^2}{dt_i \varepsilon_q} + ... + \frac{dmc_{xn}^2}{dt_i \varepsilon_{xn}} + n_{xn}\frac{dmc_q^2}{dt_i \varepsilon_q} =$$

$$= \frac{dmc_{x1}^2}{dt_i \varepsilon_{x1}} + \frac{dmc_{x2}^2}{dt_i \varepsilon_{x2}} + ... + \frac{dmc_{xn}^2}{dt_i \varepsilon_{xn}} ;$$

5D1

➤ The quantum signals of one and the same event having been acting within the quantum system of different space-times:
Space-times with different quantum speed values are equal parts of the integrated quantum system and contain the same quantum of the integrated system. In the case of one and the same event, the numbers of quantum impacted in different space-times are different $n_{x1} \neq n_{x2} \neq ... \neq n_{xn}$ but space-times contain each other. Space-times with higher number of quantum impacts contain the others, with less numbers impacted. Otherwise, there would be no common quantum impulse (quantum) impaction.
With reference to 5C7, the number of the quantum impacted depends

Ref.
5C7
Ref.
5C5

$$\text{on the value of the quantum speed: } f(c_x) = f(n_x).$$

The intensity of the event within the space-time, with reference to 5C5 is:
$$\varepsilon_x = \frac{EVENT}{dt_x} = \frac{EVENT}{dt_i}n_x$$

5D2

dt_x - is not the duration of the event here, rather the characteristic of the time system;

n_x - is the impacted by the quantum speed quantum within space-time x;

dt_i - is the time system of all quantum impulses (and *blue shift* impacts);

The intensity of the event, related to a single quantum:
$$\varepsilon_q = \frac{\varepsilon_x}{n_{xq}} = \frac{EVENT}{dt_x n_{xq}} = \frac{EVENT}{dt_i}\frac{n_x}{n_{xq}}$$

5D3

n_{xq} - is the *number of the impacted quantum* of the event – means the intensity of the signal of the event, within the quantum system of space-time x.

Here above in 5D2 and in 5D3 the value directly means the intensity.
(In elementary processes the value is reciprocal)

The purpose of this specific kind of interpretation is for having quantum related reference to characterise the event. The time count of the space-time is a valid value and the number of the impacted quantum also describes the intensity of the event within the quantum system of the given space-time.

5D2 and 5D3 give: $EVENT = (1) = \varepsilon_x dt_x = \varepsilon_q n_{xq} dt_x;$ and $\varepsilon_x = \varepsilon_q n_{xq}$

5D4

In the case of the same event just happening in the different quantum system of y:
$$EVENT = (1) = \varepsilon_y dt_y = \varepsilon_q n_{yq} dt_y; \quad \text{and} \quad \varepsilon_y = \varepsilon_q n_{yq}$$

5D5

As with reference to 5C5 $dt_x \neq dt_y$ and $n_x \neq n_y$ $\rightarrow$ $\varepsilon_x \neq \varepsilon_y$ and $n_{xq} \neq n_{yq}$

5D6

5D4, 5D5 and 5D6 mean: While the intensity of a single *quantum* ε_q is equal and the same in all space-times and corresponds to its infinite low value, *the space-time is the one which determines the number of the impacted quantum* of the quantum signal of the event. The space-time is the one which decides what the number of the quantum of the realisation of the event in that given space-time is.

This number can be changing from $\lim n_{xq} = 0$ to $\lim n_{xq} = \infty$. And these numbers of quantum impulses (quantum) contain each other. Space-times with *higher* quantum speed values contain all other quantum of space-times with less quantum speed.

There is an important note:
- the quantum system (space) is one and the same, and
- events happening within the quantum system (space) are one and the same,
- the measured duration and the intensity of the events within different space-times are the ones which are *different*!

S.
5.2.1

5.2.1. The space-time of the *Earth* surface

We live within our *space-time* on the *Earth* surface with our acting quantum speed of "light". We sense our events in line with our time flow.

These "our events" on the surface of the *Earth* might have different durations within other space-times of different quantum speed values. For us, however, only our events with the quantum speed of light on the *Earth* surface can be "registered". We cannot receive any other information, just the one provided by our space-time and time system.

Events happen simultaneously.

There is no way to participate within the same event within two or more time systems in parallel, or repeating an event within different space-times.

Each elementary process has its own quantum speed value.

Elementary quantum communication happens in line within quasi equal dt_i time system and with the passive/neutral character of the neutron process.

5E1

$$e_x = \frac{dmc_x^2}{dt_i \varepsilon_x}\left(1 - \sqrt{1 - \frac{(c_x - i_x)^2}{c_x^2}}\right)$$

If impacting the neutron process of other elementary processes, the acting electron process *blue shift* drive generates its own space-time by its quantum speed value.

The driven neutron processes will be parts of the elementary processes of the drive.

While the communicating elementary processes keep their own space-times, in long term, the space-time with higher quantum speed prevails.

Our space-time on the *Earth* surface is generated by the anti-electron process *blue shift* impact of the *plasma* and the evolution of elementary processes = *gravitation*.

Ref.
S.5.1

With reference to Section 5.1, the quantum communication of the first 30 elementary processes of the periodic table is easy, as the elementary space-times are close to each other and the communication is balanced.

In the case of significant quantum speed difference, the elementary communication is more complicated: While the communication has its certain direction and the elementary process with the higher quantum speed is the dominant one, its internal intensity capacity, sooner or later, will limit the communication and will stop it. The reason is the significant loss in its quantum potential – quantum speed value.

5.2.2. *Inflexions* of events

Inflexions are of $\Delta t = 0$.

All *inflexions,* whatever the quantum speed values of the space-times and the intensities of the events are, have *zero* time duration.

The *quantum* drive of the elementary process is $IQ = \dfrac{c_x^2}{\varepsilon_x}$;

Fig.5.2 demonstrates the differences between inflexions.

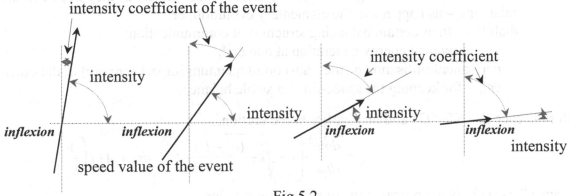

Fig.5.2

Fig.
5.2

The length of the line is proportional to the value of the quantum speed.
The intensity of the event is proportional to the value of the angle with blue arrow.
In the case of a certain *space-time*, the quantum speed is constant, but the intensities of events can vary.

5.2.3. Space-times are "distance-independent"

Separated space-times of equal quantum speed establish their common space-time and are communicating.
The proofs are summarised below:
- Space is the system of quantum impulses (quantum).
- Space-time is characterised by quantum speed, therefore, there is no difference where quantum signals of equal quantum speed are acting; there are no such categories within space-times as "far" or "close", "short" or "long distance".
- Quantum impulses (quantum) are being generated and are establishing the quantum system. The quantum system is impacted by quantum signals, variety of quantum speed values.
- While quantum signals, in conventional terms, might be impacting the quantum system "close" to, or "far" from each other, the generating space-time depends only on the speed value of the quantum communication.
- The higher the quantum speed is, the "wider," in conventional terms, is the quantum system which the quantum impact is covering.
- If there are single quantum signals acting of equal quantum speed, generated independently from each other, the chance for establishing common space-time is low, as single quantum signals loose on their intensity in quantum conflicts.

- Single quantum impacts cannot establish, in principle, neither their own nor common space-times.
- In the case of permanent generation of quantum impacts, the feeding source always covers the intensity loss, being lost in conflicts; in this way, quantum signals can keep their original intensity and quantum speed values; there is no difference where, in conventional terms, the quantum signals, establishing the common space-time, have been generating; these signals establish their common space-time "independently on the distance from each other".

These space-times
- either are communicating with other, different space-times, as they need certain balancing – as happens in the elementary evolution, or
- shall have their certain balancing structure of communication:
  - the permanent energy generation at one end;
  - permanent utilisation of the transported quantum impact (energy) at the other end – for keeping the space-time in stable balance.

Ref.
1N5
5C3
5E1

With reference to 1N5, 5C3 and 5E1 earlier in the section,

$$e_x = \frac{dmc_x^2}{dt_i \varepsilon_x}\left(1 - \sqrt{1 - \frac{(c_x - i_x)^2}{c_x^2}}\right) \quad ds_x = A \cdot f\left(\frac{1}{c_x}\right)$$

"distance," as such, is reciprocal to the quantum speed value.

➤ High quantum speed means a large number of impacted quantum (for the unit period of time), a "large" space in conventional terms; therefore, the space-time is with "slow time flow", as a huge volume of information is being transported.
➤ Low quantum speed is equivalent to a low number of impacted quantum, which corresponds to "smaller" space, with less information; therefore, it is with speeded up time flow (meaning: the communication of the same information needs more time).

Distance-independence is coming from the *inflexions* of the elementary processes (of the generation of the quantum signals).
  Space-times are cross-communicating and have been connected.

5F2

Not only by the equal time systems of all quantum signals and impacts: $\quad dt_i = \dfrac{dt_{xo}}{\sqrt{1 - \dfrac{i_x^2}{c_x^2}}}$
  but mainly by dt_{xo}, the time systems of the *inflexions*.

  The elementary *space-time matrix* contains all *space-times*.
Whatever the time systems and the *space-times* of the *inflexions* are *inflexions* happen for equally *zero* time duration in any system. But the *inflexions* of equal quantum speed values happen within the same time system of the elementary process. This way the quantum speed value determines the elementary process of the origin. And equal quantum speed values have the same origin, the same space-time.

Quantum signals with increased quantum speed might happen in space-times of less intensity. Never the opposite. Quantum signals of less intensity cannot impact the quantum system of a *space-time* with increased quantum speed and intensity. Therefore the *space-times* of higher quantum speed and increased intensity, with reference to Section 5.1, contain all other *space-times* with less quantum speed values, from infinite high to infinite low. And the quantum signals of equal quantum speed values compose the same *space-time* whatever would it be in conventional terms the distance between them.

Ref.
S.5.1

5.3
The duration and the intensity of events having been influenced by motion

1. Space-times are about the speed values of the quantum communication.
   Events with the infinite variety of intensities may happen within the space-times.

The intensity of the quantum impact of an event means a certain n_x impacted quantum for the unit period of time:

$$\frac{dn_x}{dt_x} \cdot q = \frac{dmc_x^2}{dt_i \varepsilon_x}\left(1 - \sqrt{1 - \frac{(c_x - i_x)^2}{c_x^2}}\right)$$

5G1

The drive of the event is the intensity quotient: $IQ_x = \frac{c_x^2}{\varepsilon_x}$; of the elementary process

The higher the IQ value is, the higher is n_x, the number of the quantum impacted.

5G2

At a certain equal number of quantum impacts, the higher the IQ value, the shorter is dt_x, the duration of the event within the space-time of the given quantum speed.

As $dt_i = \dfrac{dt_o}{\sqrt{1 - \dfrac{(c-i)^2}{c^2}}}$

whatever is the quantum speed value of the space-time, the time system of the quantum system is of infinite long duration: $i = \lim a\Delta t = \lim c$.

dt_o is the *inflexion* within the space-time matrix of the elementary processes.

5G3

High intensity signals (events) impact high number of quantum.
High number of impacted quantum means large "space" within the space-time with c_x as the impact is approaching high number of quantum impulses.
If the quantum impact of the event is n_x, the signal is impacting n_x quantum for the unit period of time. (The reference to the unit period of time is necessary for pointing out that this is about intensity.)
In the case of conflicts the transported signal obviously becomes interrupted, but in principle the number of the quantum is the one expressing the capacity of the impact and creating its own space-time.

The quantum impact of the event means immediate transfer of the impact by the quantum impulses in any direction. The number of the quantum participating within the transfer of the quantum signal corresponds to the intensity (the IQ value, the quotient of the speed of quantum communication and the intensity coefficient, as above) of the impact.
There is a difference between dt_i the time parameter of the quantum impulse (quantum) of the quantum system, and dt_x the time parameter of the event:

> the overall quantum system, quantum impacts and single quantum impulses have the same time system parameter = dt_i infinite long as above;

> dt_x the time parameter of the *space-time* depends on the intensity of the event, the number of the impacted quantum; its value varies from *zero* to *infinity*.

Communicating with N_x quantum in total means $N_x \cdot q$ quantum impact, with $\dfrac{dn_x}{dt_x}$

5G4

intensity of the impact. These are characterising the event.

5G5   2. An event within a certain space-time, transported by a motion of certain speed value is continuously impacting the quantum system and impacting n_{sum} (n_x and n_v) quantum,

where n_x relates to the event and n_v is the consequence of the motion.

5G6   The intensity of the event in fact has two components: $\dfrac{dn_x}{dt_x}$; and $\dfrac{dn_v}{dt_x}$

If assessing the impact of the motion, the higher is the speed of the motion, the higher is the impact within the space-time, where the motion and the event are taking place.
The higher of the speed of the motion of the event is the higher is the quantum communication of the event with the quantum system.

5G7   If taking the speed of the motion of an *event* for v, the intensity of the $\dfrac{dn_v}{dt_x}$
impact, consequence of the motion will be:

Higher speed means higher intensity, higher quantum impact for the unit period of time.
The intensity of the event, as consequence of the motion gives certain duration:

5G8 $\dfrac{N_v}{\dfrac{dn_v}{dt_x}} = f(t_x)$ - time dimension
within the denominator is the quantum number ,
with the intensity in the nominator.

The higher the speed of the motion is, the less is the resulting time; the shorter is the duration of the event within the time system of the motion.
dt_x is the time parameter of the event in its quasi static status.

The intensity of an event within the system of reference in motion:
-    The basic number of the impacted quantum of the event is the same whatever is the status (rest or motion) of the event;
The increase of the number of the impacted quantum, as consequence of the motion depends on the speed of the motion. The higher is the speed, the higher is the impacted quantum.
The intensity of the event within the system of the motion, relative to the time system, where the motion is taking place:

5G9 $\varepsilon_e = \dfrac{EVENT}{\Delta t_e}$; and $\Delta t_e = \dfrac{EVENT}{\varepsilon_e}$ The higher the speed of the motion is, the higher is the intensity of the event and less is the duration.

The time count (the duration) of the event within the space-time depends on the intensity of the event and on the speed of the motion of the event. The motion is increasing the quantum communication and the event is impacting more quantum – resulting in the increase of the intensity. Increasing intensity is equivalent to the shortening of the time flow.
The time is the relation of the event to the quantum system.

Event without motion, in static status:

$$EVENT = N_x = \frac{dn_x}{dt_x}\Delta t_x ; \text{ and}$$

$$\Delta t_x = \frac{N_x}{dn_x}dt_x$$

Event in motion:

$$EVENT = N_x = \left(\frac{dn_x}{dt_x} + \frac{dn_v}{dt_x}\right)\Delta t_v ; \text{ and}$$

$$\Delta t_{xv} = \frac{N_x}{dn_{xv}}dt_x ; \quad \text{where } \frac{dn_x}{dt_x} + \frac{dn_v}{dt_x} = \frac{dn_{xv}}{dt_x}$$

5G10

and $\Delta t_x > \Delta t_{xv}$

The result of the simplified assessment in 5G10 means, motion is increasing the intensity of the quantum communication of the event, therefore the duration of the event within the system of reference in motion will be shorter: $\Delta t_{xv} < \Delta t_x$.

Ref. 5G10

This reduced duration in 5G10 is expressed by the time count of the event in its static status. In the case the motion stops, the time, spent by the event in motion is less than the time, spent for the same event in its static status.

With the increase of the intensity of the quantum communication, with reference to 5G6, the duration of the event within the system in motion will be equal to Δt_{xv} $\dfrac{dn_x}{dt_x}$; and $\dfrac{dn_v}{dt_x}$

Ref. 5G6

Comparing it to the time system at relative rest, the intensity of the event is increasing. But the event remains the same.

The intensity

within the time system of quasi static status: $\varepsilon_x = \dfrac{EVENT}{\Delta t_x}$;   within the time system in motion: $\varepsilon_{xv} = \dfrac{EVENT}{\Delta t_{xv}}$;   5H1

And: $\varepsilon_x \Delta t_x = \varepsilon_{xv} \Delta t_{xv} = 1 = (EVENT)$; $\varepsilon_x < \varepsilon_{xv}$ and $\Delta t_x > \Delta t_{xv}$ 5H2

For assessing the relation of the durations and the capacities of the event in motion relative to the quasi static system the approach is simple:

Ref. 5G10

The duration of the event is equal to the number of the total quantum impact: N_x

With reference to 5G10

$$\frac{dn_x}{dt_x}\Delta t_x = N_x = EVENT;\qquad \text{and}\quad \left(\frac{dn_x}{dt_x}+\frac{dn_v}{dt_x}\right)\Delta t_{xv} = N_x = EVENT \qquad 5H3$$

The event is the same whatever is the intensity of the quantum communication.

$\dfrac{dn_v}{dt_x}$ means the intensity surplus, consequence of the motion, expressed by the time count of the system at relative rest.

The intensities are:

$$\varepsilon_{xv} = \frac{dn_x}{dt_x}+\frac{dn_v}{dt_x} = \left(\varepsilon_x + \frac{dn_v}{dt_x}\right);\quad \text{and}\quad \left(\varepsilon_x + \frac{dn_v}{dt_x}\right)\Delta t_{xv} = N_x \qquad 5I1$$

$$\frac{\varepsilon_x}{N_x}+\frac{\varepsilon_v}{N_x} = \frac{1}{\Delta t_{xv}};\qquad \text{where } \varepsilon_v = \frac{dn_v}{dt_x} \text{ and taking for } \varepsilon_v = \eta\varepsilon_x; \qquad 5I2$$

$$\frac{\varepsilon_x}{N_x}+\eta\frac{\varepsilon_x}{N_x} = \frac{\varepsilon_x(1+\eta)}{N_x} = \frac{\varepsilon_{xv}}{N_x};\quad \frac{\varepsilon_x}{N_x}(1+\eta) = \frac{\varepsilon_{xv}}{N_x};\quad \varepsilon_x(1+\eta) = \varepsilon_{xv} \qquad 5I3$$

$N_x = const$

$$\varepsilon_x = \varepsilon_{xv}\frac{1}{1+\eta};\quad \Delta t_x = \frac{\varepsilon_{xv}\Delta t_{xv}}{\varepsilon_{xv}}(1+\eta) = \Delta t_{xv}(1+\eta);\quad \text{and } \Delta t_{xv} = \frac{\Delta t_x}{1+\eta} \qquad 5I4$$

5I4 gives the duration of the event happening in the system of reference in motion.

If any time moment the motion stops, the time, spent by the event in motion is less than the time, spent by the same event at rest within the space-time with quantum speed c_x. This does not mean there are in this case two time counts within the system of reference at rest. The time is passing in line with the system of reference at rest, but with less time value (duration) used by the event in motion.

This directly means the remaining surplus in capacities!

The event in motion saves time and energy.

The acting intensity in motion is higher, but the *"contribution"* of the event itself, the factual (dn_x/dt_x) intensity/energy input remains the same as it is within the static status.

Ref
5G10
5H2
5I4

Compering identical results,

The time spent in static status is: Δt_x	And with reference to 5G10, 5H2 and 5I4
The time spent as result of the motion: Δt_{xv}	$\Delta t_{xv} < \Delta t_x$

The reduction of the duration of the event can also be proven by the work formula of the acceleration as well.

5J1

If the event is *taken*, its work intensity formula is: $w_{xv} = \dfrac{dmc_x^2}{dt_x}\left(\dfrac{1}{\sqrt{1-\dfrac{v^2}{c_x^2}}} - 1\right)$;	In the contrary to the case when the event itself is the drive: $w_v = \dfrac{dmc_x^2}{dt_x}\left(1 - \sqrt{1-\dfrac{v^2}{c_x^2}}\right)$

5J2   The time count within the system of reference in motion in this case is: $dt_{xv} = dt_x\sqrt{1-\dfrac{v^2}{c_x^2}}$

capacity reserve

At the time moment the event at relative rest status is completed as its intensity/energy reserves has been expired, (with ε_x intensity and time duration Δt_x).	The similar event within the system in motion has already been completed, with still remaining intensity/energy capacities. As the duration was Δt_{xv} and the intensity for this period also was ε_x.

5J3   If the event in motion stops and $\dfrac{dn_x}{dt_x} = 0$, $\dfrac{dn_v}{dt_x}$ makes no sense anymore.

If an event having been in motion, returning back into the system of reference at rest again, it continuous with time system of the rest, but from less "aged" status.

Ref.
5B2

The case of different *space-times* with different speed value of quantum communication is simple. With reference to 5B2 the *space-time* with higher quantum speed has slower time flow and higher intensity.

S.
5.4

5.4

Infinite low intensity is turning into infinite high value

The *plasma* is the natural consequence of the accumulating infinite large number of *Hydrogen* processes, the endpoints of elementary evolution. With the last elementary process transforming into *Hydrogen* process, the space-time is reaching its infinite low intensity, corresponding in conventional terms to being infinite small. The collapse with infinite high intensity is unavoidable.

The increased to infinite high intensity of the collapse of the neutron processes of infinite low intensity is consequence of the accumulated infinite high number of *Hydrogen* processes and the infinite small space-time.

6

Energy generation by the electron process *blue shift* conflict
of the *Hydrogen* process and *gravitation*

The neutron process of the *Hydrogen* process is of infinite low intensity. The reason is the intensity of the electron process *blue shift* drive, which is also of infinite low intensity. The elementary equation in absolute terms is:

$$\frac{dmc^2}{dt_p \varepsilon_p}\left(1 - \sqrt{1 - \frac{i^2}{c^2}}\right) = \xi \frac{dmc^2}{dt_n \varepsilon_n}\sqrt{1 - \frac{(c-i)^2}{c^2}}\left(\sqrt{1 - \frac{i^2}{c^2}} - 1\right)$$

6A1

The natural intensity of the electron process of the *Hydrogen* process is:

$$\frac{dmc_H^2}{dt_i \varepsilon_H}\left(1 - \sqrt{1 - \frac{(c_H - i_H)^2}{c_H^2}}\right)$$

where c_H in fact is of infinite low value $\lim c_H = 0$

Ref.
S.1.1
S.1.2
6A2

Therefore, the IQ of the *Hydrogen* process (result of *plasma*) is: $\lim IQ_H = 0$.
The intensity of the electron process on the surface of the *Earth* with the impact of *gravitation* still remains of infinite low value, as below in 6A3. But the quantum speed is

$$\frac{dmc^2}{dt_i \varepsilon_H}\left(1 - \sqrt{1 - \frac{(c-i)^2}{c^2}}\right);$$

$c = 299{,}792$ km/sec, which corresponds to the quantum speed of the (at the) *Earth* surface the speeding by the acceleration of the *Earth* surface.

6A3

Since the intensity coefficient of the *Hydrogen* process is $\lim \varepsilon_H = \propto$, even the increased quantum speed does not change its IQ value

$$\lim IQ = \lim \frac{c^2}{\varepsilon_H} = 0$$

6A4

The time system of the quantum communication of the *Hydrogen* process remains even at the increased quantum speed of $c = 299{,}792$ km/sec equal to the time system of its original value, as at $i_H = \lim a\Delta t = c_H$

$$dt_i = \frac{dt_o}{\sqrt{1 - \frac{i^2}{c^2}}} = \frac{dt_o}{\sqrt{1 - \frac{i_H^2}{c_H^2}}}$$

6A5

Ref.
S.5.1

The time system on the surface of the *Earth* dt_{Life} in the case of "normal/usual" events ($v \ll c$) corresponds to the time system of the *Carbon* process – in line with the space-time matrix.

$$dt_{Life} = \frac{dt_C}{\sqrt{1 - \frac{v^2}{c^2}}}$$

6A6

The time system of the first 30 elementary processes of the periodic table is quasi equal to the time system of the *Carbon* process.

6.1
Plasma

The *Hydrogen* process is at one end. *Plasma* is at the other.
The intensity of the electron process of the *plasma* is of infinite high value. The intensity coefficient of the *plasma*, in line with this, is $\lim \varepsilon_{pl} = 0$:

$$\lim \varepsilon_{pl} = \lim \frac{dt_n}{dt_p} = \lim \frac{\varepsilon_p}{\varepsilon_n}\sqrt{1 - \frac{(c-i)^2}{c^2}} = 0; \quad \text{and the } IQ \text{ value:} \quad \lim IQ = \lim \frac{c_{pl}^2}{\varepsilon_{pl}} = \propto$$

6B1

From 6B1 follows:

- the infinite high intensity of the electron process results in infinite short process time of the neutron process of the *plasma* – the duration of the neutron process is approaching zero: $\lim \Delta t_n^{pl} = 0$;

- *plasma* is with two key parameters: $\lim c_{pl} = \infty$ and $\lim \varepsilon_{pl} = 0$, the quantum speed and the intensity of the electron process *blue shift* impact are of infinite high values;

- the infinite high intensity and *IQ* value of the electron process results in *blue shift* conflict of infinite high intensity;

- electron process *blue shift* conflict is equivalent to temperature increase;

- if the conflict is of infinite high intensity, the temperature is also of infinite high value;

- neutron processes are neutral – driven by electron processes, products of the proton process; therefore, the correct formulation is: the developed by the proton process electron process *blue shift* impact of the *plasma* state drives the neutron process by infinite high intensity;

- this also means that this is not about a specific proton process, capable of developing an electron process impact of infinite high intensity, but rather: $\lim dt_n = 0$!

- [contrary to the *Hydrogen* status, where the intensity of the *blue shift* impact of the electron process, the drive of the neutron process, is of infinite low intensity and infinite long duration; but in that certain case $\lim dt_n^H = \infty$ and $\lim \varepsilon_H = \infty$];

- relativity is the key;

- [in the case of the *Hydrogen* process the gaseous state and the $c = 299,792$ *km/sec* quantum speed is caused by *gravitation* (!)];

- *plasma* is cooled by *gravitation*, taking off *blue shift* impact intensity, as the *Quantum Membrane* of the plasma is expanding in space and time – working out (resolving) the conflict and establishing the balance;

- intensity relations have been regulated and established by space and time (creating elementary processes): expanded space results in intensity decrease and time increase $e_x = \dfrac{e_o}{x \varepsilon_x}$ and $\varepsilon_x = \dfrac{\Delta t_x}{\Delta t_o}$

Ref
8E2

- on the way from *plasma* to the *Earth* surface, the intensity of the electron process *blue shift* impact of elementary processes is decreasing (the value of ε is growing) with the increase of the time count – the gradient of the intensity of the *blue shift* impact is decreasing;

- approaching *plasma* from the cooled by *gravitation* side, with the increase of the intensity of the electron process and the *blue shift* conflict, elementary structures become more and more conflicting as the temperature, the indicator of the conflict, is increasing;

- [*fire* is of infinite high *blue shift* conflict at the actual quantum speed and the intensity parameters of the *quantum system of reference* of the event; ($c = 299,792$ km/sec – measured on the surface of the *Earth* BUT $\varepsilon_E = 1$, taken as balanced state.)]

So, there are

➤ the *plasma* with electron process *blue shift* impact, conflict and quantum speed, the drive of the neutron process, of infinite high intensity, approaching $\lim \Delta t_{pl} = 0$ at one end, and

➤ the *Hydrogen* process, with electron process *blue shift* drive and quantum speed, both of infinite low intensities, with $\lim \Delta t_H = \infty$ at the other.

The comparison of the intensities of the proton and the neutron processes is relativistic, as the neutron process is the driven and is the passive one, while the proton process is always the intensity source of the change of the energy/mass status. Therefore, the neutron process cannot be "more or less intensive" on its own. It is the result of the elementary quantum communication with clearly defined objectives of the balance.

The status of the elementary processes decides what are the *relation* of the intensities of the proton and neutron processes (the *Strong Interrelation*) driven by the third, the electron process (the *Weak Interrelation*).

Plasma	*Hydrogen*
The neutron process at the *plasma* state is driven by electron process of infinite high intensity, causing conflict of infinite high intensity and electron process *blue shift* impact of infinite high intensity;	The neutron process at *Hydrogen* state is driven by infinite low intensity value, by an electron process *blue shift* impact, lasting for infinity.

The *inflexion* point at *plasma* is of infinite high intensity – initiating anti-neutron process expansion of infinite high intensity and anti-electron process with infinite large surplus. The *inflexion* of the *plasma* and the *Hydrogen* process is one and the same. The infinite low intensity of the *Hydrogen* process turns into its opposite: into infinite high intensity.

Earth is part of the overall elementary balance.
The *Earth* "process" – by the acceleration for infinite time (*gravitation*), and being part and subject (with reference to Section 1) to the elementary evolution – is working out the conflict initiated by the increased intensity of the *plasma* status.
Our space-time on the *Earth* surface exists in parallel with the space-times of the *plasma* of infinite high intensity and the *Hydrogen* process of infinite low intensity.

Quantum communication has its general time formula	The intensity of the electron process is function of the *IQ* of the element	Time count also depends on the speed of the motion of the elementary process	
$$dt_i = \dfrac{dt_o}{\sqrt{1-\dfrac{i_x^2}{c_x^2}}}$$	$$e_x = \dfrac{dmc_x^2}{dt_i \varepsilon_x}\left(1-\sqrt{1-\dfrac{(c_x-i_x)^2}{c_x^2}}\right)$$	$$dt_v = \dfrac{dt_o}{\sqrt{1-\dfrac{v^2}{c_x^2}}}$$	6B2 6B3 6B4 Ref. 5.1

dt_o is taken as the time system of relative rest, but in fact it is representing the time system of the *inflexion points* in line with the overall space-time matrix of elementary evolution .
Looking at the time formula in 6B4:
- both the *speeding up*, the speed increase of v the motion of the system of reference and the *increase* of c_x, the quantum speed result in the shortening of the time flow and in the *increase of intensity*;
- both $v=0$ and $\lim c_x = \infty$ are resulting in the status of approaching rest;
- the difference between $v=0$ the speed value of the system of reference and $\lim c_x = \infty$ the speed value of quantum communication is that
  - $\lim c_x = \infty$ leads to the status of *relative* rest = to the *inflexion point*;
  - $v=0$ to the status of rest within (or *relative* to) the system of reference of the assessment.

Plasma state is: $\lim c_{pl} = \infty$ and $\lim \varepsilon_{pl} = 0$

The *Earth* surface with $c_E = c = 299{,}792$ km/sec and $\varepsilon_E = 1$ is the result of the expanding acceleration of *plasma*, which is the result of the evolution of the elementary processes.

Hydrogen process state is: $\lim c_H = 0$ and $\lim \varepsilon_H = \infty$

6C1   Time relations in any *Quantum Systems of Reference*, taken at relative rest are controlled by c_x, the *quantum speed* of the system of reference:

$$dt_{vx} = \frac{dt_{ox}}{\sqrt{1 - \dfrac{v_x^2}{c_x^2}}}$$

6C2   *Gravitation* for the *plasma* is the result of the elementary evolution: $\lim g_{pl} = \infty$

6C3   *Gravitation* for the *Hydrogen* process is the drive increasing its quantum speed: $\lim g_H = 0$

6C4   The time system of the electron process *blue shift* impact in any quantum system is:

$$dt_{ix} = \frac{dt_{ox}}{\sqrt{1 - \dfrac{i_x^2}{c_x^2}}} = \frac{dt_{ox}}{\sqrt{1 - \dfrac{\lim g_x \Delta t}{g_x \Delta t}}}$$

6C5   The *IQ* value is the one in fact establishing the time system of the *Quantum System of Reference*. $IQ = \dfrac{c_x^2}{\varepsilon_x}$

6C6   *IQ* for *plasma* is: $\lim IQ = \infty$; for the *Hydrogen* process is $\lim IQ_H = 0$

IQ values for the *Earth* surface and for all of the elementary processes of the periodic table are given in Table 1.1 of Section 1.

Elementary processes are about the balance of proton and neutron processes, about the relation of their intensities, established by ε_x the intensity of the electron process *blue shift*

Ref.   impact. The increase of the intensity of the elementary process initiated by elementary
6B5   quantum communication or by external impact of any kind results in conflict.
6C2   The consequence of the conflict is heat generation.

Without controlling it and cooling it by *gravitation,* the conflict results in *fire* = electron process *blue shift* conflict of infinite high intensity – at the given quantum speed.

Fire is annihilation and destruction.

BUT *destruction* has not been encoded into the evolution of the elementary world, the life process of the *Earth.*

Nature is about resolving conflicts.

Elementary conflicts are the drives of the progress, the drives of the natural *plasma-Hydrogen* "evolution" line. Elementary communication is for establishing the balance again and again.

Plasma is NOT *fire!* – *Earth plasma* corresponds to an *inflexion* point.

- with the time count of *zero*: $\lim dt_o = 0$

- with the quantum speed of *infinite* high value: $\lim c_{pl} = \infty$

- with the intensity of the electron process *blue shift* impact of *infinite* high value: $\lim \varepsilon_{pl} = 0$

Passing the *inflexion* point of the *plasma* all parameters are changing: $\lim IQ_{pl} = \infty$ is approaching $\lim IQ_H = 0$ with $\lim \varepsilon_H = \infty$ and $\lim c_H = 0$ of the *Hydrogen* process – going through all steps of the elementary evolution, going through all statuses of the existence of the elementary processes.

6.2
After *Plasma* approaching finally to the *Hydrogen* process

With reference to 6B1,

$$\lim \varepsilon_{pl} = \lim \frac{dt_n}{dt_p} = \lim \frac{\varepsilon_p}{\varepsilon_n} \sqrt{1 - \frac{(c-i)^2}{c^2}} = 0; \quad \text{and} \quad \lim IQ = \lim \frac{c_{pl}^2}{\varepsilon_{pl}} = \infty$$

Expanding *plasma* means being cooled by *gravitation*.
The intensity of the electron process is partially leaving, taken away, the *IQ* value is decreasing.
With the expansion, approaching external surface of the *Earth*:
- the intensities of the electron processes of the elementary processes become less and less in value towards the *Earth* surface, taken $\varepsilon_E = 1$;

  [$\varepsilon_E = 1$ means the equal intensities of the proton and the neutron processes. Electron process *blue shift* conflicts mean gaseous or/and liquid states; minerals have intensity coefficients, representing balanced elementary statuses.]
- the speed of quantum communication is less and less approaching $c = 299,792$ km/sec measured on the *Earth* surface.

The intensity characteristics, the quantum speed and *IQ* values of elementary processes are given in Table 1.1 of Section 1.2.

All elementary processes have their own quantum speed.
Quantum communication in natural circumstances happens at the quantum speed of the elementary process. Minerals and mix of elements represent the integrated quantum speed value of the elementary components.
In the case of water, this is about $c_{water} = 225,000$ km/sec, proven by experiment.

Quantum communication of elementary processes is harmonisation of the intensities of the quantum systems:
Elements with neutron process dominancy have electron process *blue shift* impact of increased intensity. The electron process *blue shift* impact of proton process dominant elementary processes is of less intensity.

Quantum communication is balancing the intensities of the *Quantum Membranes*.
One side is losing, while the other is gaining on the *blue shift* impact intensity potential.
The elementary composition becomes stable, with integrated intensity of the *Quantum Membrane*, with harmonised common quantum speed value.
This is the reason for the reduced quantum speed values of *water* and *Hydrocarbons*.
The infinite low value of the quantum speed of the natural *Hydrogen* process reduces the integrated acting speed values.

Gas status above the *Earth* surface is with the quantum speed of *gravitation*.
Liquid status is with electron process *blue shift* surplus. In the case of natural *Hydrogen* process, with reference to 6E1, the natural *Hydrogen* process means reduced quantum speed.
The solid status of crystals and minerals is with increased quantum speed.

The quantum speed of some *Hydrocarbons*: (km/sec)		While gases as expected, at the quantum speed of *gravitation* are: (km/sec)			
Acetone	221,506	Air	299,917	Krypton	299,871
Butane	215,444	Carbon Dioxide	299,866	Methane	299,866
Ethanol	220,997	Ammonia	299,887	Neon	299,980
Methanol	228,230	Helium	299,989	Nitrogen	299,910
Glycerol	204,578	Hydrogen	299,958	Oxygen	299,918

Data will harmonise with Table 1.1 of Section 1.2.   Source: *http/:www.refractiveindex.info/*

Table 6.1

Table 6.1

Proton process cover, electron process drive and neutron collapse cannot be the events of the same elementary cycle. The reason is the sequence of events:

Proton process of an elementary cycle gives cover to the neutron collapse of the same cycle, but of an earlier sequence. Proton process continues as electron process drive, driving the neutron process of the same cycle, but of an earlier sequence passing already proton and electron processes; while the proton process cover is given by the following sequence, the actual electron process is, in fact, from the previous proton process.

Electron process continues as neutron collapse, driven by the electron process of the next cycle, the proton process of the previous sequence; the proton process of the (new) cycle ensures the cover.

The intensity of the neutron collapse at the *inflexion point* is the load of the expansion of the anti-neutron process. It combines cycles.

The *entropy* is the constant loss of the mass/energy intensity of each cycle. This is the drive of the sequence of elementary processes. The infinite high mass/energy intensity concentration of *plasma* transforms into the infinite low mass/energy intensity value of the *Hydrogen* process.

The following table-diagram demonstrates the cycle and the sequence of elementary processes.

Horizontal: the elementary process, the integration of three events
Vertical: the development of events in sequences.

proton	electron	neutron	inflexion
electron	neutron	inflexion	proton
neutron	inflexion	proton	electron
inflexion	proton	electron	neutron

Table 6.2

Table 6.2

As the result of *Earth gravitation*, the chain of elementary transformation, the cooling of *plasma* at a certain stage results in *Helium* process, the last elementary process with full cycle. As the result of the overall process, the cycle is losing on mass/energy intensity.

At the end of the *Helium* process, the mass/energy potential of the electron process of the *Helium* cycle will be so weak that the collapse of the neutron process becomes of infinite low intensity. In this way, from one direction the process continues, as *gravitation*, the cooling of *plasma* continues, in the other direction, however, after the last completed *inflexion* of the *Helium* process, the elementary process becomes of infinite low intensity and infinite long duration.

This resulting elementary process is the *Hydrogen* process.

The intensity of the electron process drive of the neutron process is so weak that it results in neutron process of infinite low measured value, in practical terms this is taken as *zero*.

Quantum speed is $\lim c_H = 0$; the intensity of the electron process is of infinite low value, the intensity coefficient of the electron process is $\lim \varepsilon = \infty$; $\lim IQ = 0$.

Helium is with proton process intensity of 2.0145 (proton weight) and neutron process intensity of 1.9869 (neutron weight). The intensity of the electron process is: 1.01389.

The proton process intensity of the *Hydrogen* process is $p = 1.0072$; the intensity impact of the neutron process is $\lim n = 0$.

[This is the reason for the H_2 character of the *Hydrogen* process ($2 \cdot 1.0072 \approx 2.0145$) – fluency with the *Helium* process.]

The reason for the high *quantum communication potential* of the *Hydrogen* process is its neutron process, inviting electron process *blue shift* impacts of other elementary processes with capacity, intensity and surplus for collapsing it. Neutron processes are neutral, can be driven and covered by the electron and proton processes of other elementary processes.

Quantum communication		Meaning:	
happens at equal time system of quantum communication but at quantum speed of quasi *integrated* value.	$dt_i = \dfrac{dt_o}{\sqrt{1 - \dfrac{i_x^2}{c_x^2}}}$	the contributing elementary processes communicate at similar time system, even having different (their own) quantum speed.	6E2

In this way, the permanent electron process *blue shift* surplus of the *Hydrogen* process provides active elementary drive capability to the *Hydrogen* process in elementary communication at the integrated quantum speed of the communication.

As neutron processes are taken by the communicating elementary processes, the neutron process collapse demand of the *Hydrogen* process with $\lim IQ = 0$ value disappears.

The key is that the neutron processes are neutral and, in this way, are driven and covered by the electron and the proton processes of other elementary processes – resulting in the elementary process of the drive.

In this way, the infinite weak intensity potential of the proton process of the *Hydrogen*

$$\frac{dt_n}{dt_p} = \varepsilon_e = \frac{\varepsilon_p}{\varepsilon_n} \sqrt{1 - \frac{(c-i)^2}{c^2}}$$

6E3

– as the original reason of the infinite low intensity of the neutron collapse becomes free from covering the neutron collapse demand.

When dividing water to *Oxygen* and *Hydrogen* processes *by electricity*, the artificially generated electron process *blue shift* conflict separates the elementary processes, at the quantum speed of the *Earth* surface.

In this way, the quantum speed of the *Hydrogen* process remains $c = 299,792$ km/sec, but the intensity potential of the neutron process of the *Hydrogen* process, characteristic of the elementary process, still remains at infinite low level. This technology gives the opportunity for having *Hydrogen* process at the quantum speed of *gravitation*.

S.
6.3

6.3
Acceleration of the *Hydrogen* process

The use of the *internal* mass/energy potential of the system of reference of the elementary process for speeding up to speed v also speeds up the time flow and decreases the intensity of the process:

6F1
$$w = \frac{dmc^2}{dt_o}\left(1 - \sqrt{1 - \frac{v^2}{c^2}}\right)$$

The mass/energy potential has been lost, which also means the lengthening of the time flow and decreasing intensity.

In the case of the acceleration of the elementary system for the count of an *external* energy source, the speeding up energy is coming from the external system of reference in relative rest.

6F2
$$w = \frac{dmc^2}{dt_o}\left(\frac{1}{\sqrt{1 - \frac{v^2}{c^2}}} - 1\right)$$

The speeding up by the use of *external* mass/energy source extends the energy/mass potential of the process. It is equivalent to increasing the intensity. The speeding up slows down the time flow.

The system of reference is in motion with speed v, but this speed value v corresponds to the status of relativistic rest if the motion is driven from an external source, as none of its own mass/energy capacity is used for the drive. External system is the only one spending work/energy for the speeding up.

Decreased time count means the intensity of the system of reference in motion is *increased*!

6F3
$$e_e = \frac{dmc^2}{dt_i \varepsilon_H \sqrt{1 - \frac{v^2}{c^2}}}\left(1 - \sqrt{1 - \frac{(c-i)^2}{c^2}}\right)$$

The "acting" intensity of the electron process is increasing.

The speeding up of the elementary process of the *Hydrogen* process on the surface of the *Earth*, having been found in gaseous state with quantum speed value $c = 299{,}792$ km/sec, results in the generation of increased *blue shift* conflict with the *blue shift* impact gravitation

The time system of all electron processes is dt_i - therefore, the quantum communication is free.

The complete *Hydrogen* process to be accelerated is:

6G1

The proton process:	The neutron process:	The electron process:
$\dfrac{dmc^2}{dt_p}\left(1 - \sqrt{1 - \dfrac{i^2}{c^2}}\right)$	$\dfrac{dmc^2}{dt_n}\sqrt{1 - \dfrac{(c-i)^2}{c^2}}\left(\sqrt{1 - \dfrac{i^2}{c^2}} - 1\right)$	$\dfrac{dmc^2}{dt_i \varepsilon_H}\left(1 - \sqrt{1 - \dfrac{(c-i)^2}{c^2}}\right)$

The energy intensity of the acceleration is:

6G2
$$e_{acc} = \frac{dmc^2}{dt_p\sqrt{1 - \frac{v^2}{c^2}}}\left(1 - \sqrt{1 - \frac{i^2}{c^2}}\right) + \frac{dmc^2}{dt_n\sqrt{1 - \frac{v^2}{c^2}}}\sqrt{1 - \frac{(c-i)^2}{c^2}}\left(\sqrt{1 - \frac{i^2}{c^2}} - 1\right) +$$

$$+ \frac{dmc^2}{dt_i \varepsilon_H \sqrt{1 - \frac{v^2}{c^2}}}\left(1 - \sqrt{1 - \frac{(c-i)^2}{c^2}}\right)$$

The signal of the conflict is the increasing temperature of the acceleration.
The acceleration process needs constant cooling!
The acceleration of the *Hydrogen* process generates not just heat, but also electron processes in surplus for direct use!

6.4.
Utilisation of the benefit

S.
6.4

The acceleration of the *Hydrogen* process with reference to the previous Section 6.3 results in increased *blue shift* conflict with the *blue shift* (quantum) impact of *gravitation*. The higher the speed of the acceleration is, the higher is the conflict.

In order to utilise the benefit of the conflict, the technology of the acceleration needs a certain circular channel with electromagnets on both sides of the casing of the channel for accelerating the *Hydrogen* process inside.

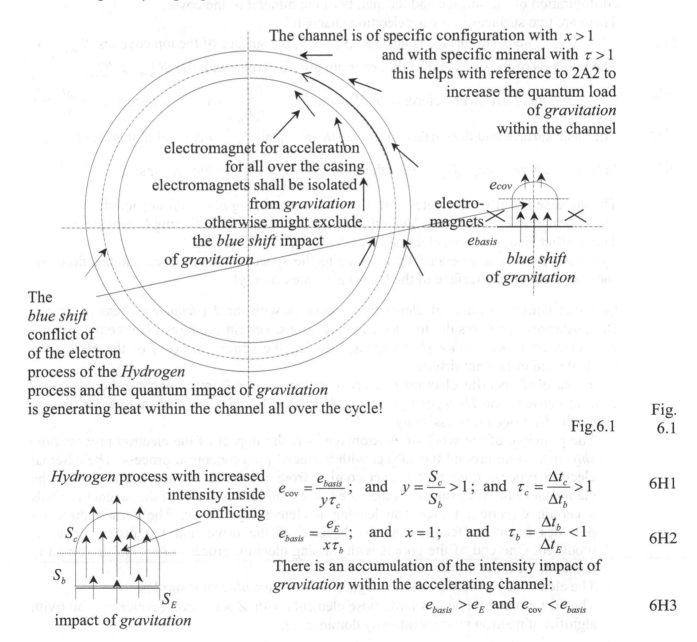

The channel is of specific configuration with $x > 1$
and with specific mineral with $\tau > 1$
this helps with reference to 2A2 to
increase the quantum load
of *gravitation*
within the channel

electromagnet for acceleration
for all over the casing
electromagnets shall be isolated
from *gravitation*
otherwise might exclude
the *blue shift* impact
of *gravitation*

electro-
magnets

e_{cov}

e_{basis}

blue shift
of *gravitation*

The
blue shift
conflict of
of the electron
process of the *Hydrogen*
process and the quantum impact of *gravitation*
is generating heat within the channel all over the cycle!

Fig.6.1

Fig.
6.1

Hydrogen process with increased
intensity inside
conflicting

S_c

S_b

S_E

impact of *gravitation*

$$e_{cov} = \frac{e_{basis}}{y\tau_c}; \quad \text{and} \quad y = \frac{S_c}{S_b} > 1; \quad \text{and} \quad \tau_c = \frac{\Delta t_c}{\Delta t_b} > 1 \qquad \text{6H1}$$

$$e_{basis} = \frac{e_E}{x\tau_b}; \quad \text{and} \quad x = 1; \quad \text{and} \quad \tau_b = \frac{\Delta t_b}{\Delta t_E} < 1 \qquad \text{6H2}$$

There is an accumulation of the intensity impact of *gravitation* within the accelerating channel:

$$e_{basis} > e_E \quad \text{and} \quad e_{cov} < e_{basis} \qquad \text{6H3}$$

There is *blue shift* impact caused by *gravitation* within the channel with *Hydrogen* process. The more the speed of the acceleration is, the more is the intensity of the electron process *blue shift* impact of the *Hydrogen* process. The more this impact is, the more intensive is the *blue shift* conflict between the *Hydrogen* process and *gravitation*.

The increasing intensity of the electron process does not change the characteristics of the *Hydrogen* process. The intensity of the neutron process still remains of infinite low value.

With the acceleration on, the increased intensity of the electron process *blue shift* impact of the *Hydrogen* process will be in conflict with *gravitation* anyway, the conflict, however, can also be increased, by increasing the intensity of the *gravitation* impact within the channel.

The way of increasing the intensity of the *gravitation* impact is to construct the acceleration channel with a certain configuration, as shown on Fig.6.1.

The intensity of the exiting the channel quantum impact of *gravitation* depends on the configuration of the surface and the quality of the mineral of the cover.

There are two surfaces in the acceleration channel:

6H4   The basic surface is taken as $S_{basis} = 1$, the surface of the top cover as $S_{cover} > 1$.

The minerals of the basis and of the cover are also different, with the $IQ_{basis} > IQ_{cov}$

6H5   therefore, with reference to 2B2 and 2A9 $\dfrac{IQ_{cov}}{IQ_{basis}} < 1$;   it means: $\tau_c = \dfrac{\Delta t_c}{\Delta t_b} > 1$

6H6   The basic surface and the surface of the *Earth* are equal: $S_{basis} = S_E$ and in this way: $x = 1$

6H7   In this case, however: $IQ_{basis} > IQ_E$, therefore $\dfrac{IQ_{basis}}{IQ_E} > 1$;   which means: $\tau_b = \dfrac{\Delta t_b}{\Delta t_E} < 1$

The increased conflict generates heat. Cooling means taking *blue shift* impact off.

The cooling medium is taking *blue shift* impact and becomes conflicting (warmed up).

The conflict also generates electron process surplus for direct use!

Hydrogen process accelerated up to close to the speed of quantum communication (the speed of light on the surface of the *Earth*) generates energy!

Quantum communication of elementary processes with the *Hydrogen* process in natural circumstances either results in water, hydrocarbons, certain hydrates of minerals, mainly with electron process *blue shift* surplus, utilising the neutron process of the *Hydrogen,* available and in fact not driven.

In the case of specific electron process quantum impact from a magnet, the *blue shift* conflict between the *Hydrogen* process and the directed *blue shift* impact of the magnet drives the *Hydrogen* process away.

> The principle of the work of electromagnets is the impact of the electron process *blue shift* flow from around the subject with balanced proton-neutron process. The external electron process *blue shift* impact coming from the wires of the magnet directs the elementary *blue shift* impact of the core (for example, iron-core) of the magnet towards a certain direction, rather than leaving it elementary-chaotic. The certain direction provides magnetic features, as the balance of the drive and the proton cover is modified. One end of the core is with missing electron process *blue shift* drive and in demand, the other is with surplus.
>
> The elementary processes of the magnets do not have *blue shift* surplus.
>
> The best magnetic features have those elements with $Z > 1.1$ event concentration (with significant neutron process intensity dominance).

The *Hydrogen* process on the surface of the *Earth* is in electron process *blue shift* conflict. This status has been proven by its gaseous state.

The directed electron process *blue shift* impact flow caused by the electricity within the wires of the electromagnets alongside the casing of the accelerating channel forces the electron process of the *Hydrogen* process into a certain direction.

Electromagnets must be isolated from the *blue shift* quantum impact of *gravitation*. Otherwise, the conflict is developing at the electromagnets rather than the *Hydrogen* process. For isolation purposes, it is best to use minerals with elementary processes with high electron process (and neutron process) intensities.

Is this about moving the complete elementary process of the *Hydrogen* including the electron process *blue shift* impact as well?
Yes, the speed increase is the key.
If the electron process *blue shift* impact from the external source would be of unlimited high intensity, capable of increasing the intensity in a single step, there would be no need for circulating the *Hydrogen* process and for reaching speed increase for approaching the needed intensity.
The *blue shift* impact from the electromagnets is moving the *Hydrogen* process within the accelerating channel. Examples of accelerators for other purposes clearly prove it.
The proton and neutron processes of the *Hydrogen* process are also forced to move.
The gaseous state of the *Hydrogen* process is the result of the electron process *blue shift* conflict. This conflict is still acting and supplemented by new ones, forcing the flow in one certain direction and *gravitation*, generating heat.

There could be a question why use the *Hydrogen* process? Why not other elementary processes in gaseous state? When accelerating *Oxygen* process, the acceleration would even need less speed value?
The answer is easy: with the increase of the intensity of the electron process, *Oxygen* process (and also the other elementary processes) would produce solidified status, because of the increasing intensity of the neutron process.
Only *Hydrogen* can be used. The intensity increase of the electron process, the result of acceleration, will not change the gaseous state.

6.5
The work value of the speeding up and the benefit of the conflict
with *gravitation*

S.
6.5

The *work value* of the speeding up, with reference to the description in 6G2:
In the case of elementary structures the mass intensity values $\dfrac{dmc^2}{dt_p}$ and $\dfrac{dmc^2}{dt_n}$ of the proton and the neutron processes are equal to

Ref.
6G2
6I1

If we want to calculate the work intensity value of the acceleration, with reference to 6G2 it needs to be

Ref
6G2
6I2

$$w_{acc} = \frac{dmc^2}{dt_p}\left(\frac{1}{\sqrt{1-\frac{v^2}{c^2}}}-1\right) + \frac{dmc^2}{dt_n}\left(\frac{1}{\sqrt{1-\frac{v^2}{c^2}}}-1\right) + \frac{dmc^2}{dt_i \varepsilon_H}\left(\frac{1}{\sqrt{1-\frac{v^2}{c^2}}}-1\right);$$

The *minus* and *plus* prefixes mean the directions of the change rather than the values of the calculated mass impact. The expansion and the collapse are constant and the intensities in 6I1 are constantly re-establishing.

6I3   From the characteristics of the *Hydrogen* process and the time system of the electron process, it follows that

$$\lim \frac{dmc^2}{dt_n} = 0; \quad \text{and} \quad \lim \frac{dmc^2}{dt_i \varepsilon_H} = 0;$$

For the electron process there are even two reasons:

6I4   1: the intensity coefficient for the *Hydrogen* is $\lim \varepsilon_H = \infty$

6I5   2: the time system $\lim dt_i = \dfrac{dt_o}{\sqrt{1 - (i_H^2 / c_H^2)}} = \infty$

Therefore, the simplified work formula of the intensity of the acceleration is:

6I6
$$w_{acc} = \underbrace{\frac{dmc^2}{dt_p \sqrt{1 - \dfrac{v^2}{c^2}}}}_{4} - \underbrace{\frac{dmc^2}{dt_p}}_{1} + \underbrace{\frac{dmc^2}{dt_n \sqrt{1 - \dfrac{v^2}{c^2}}}}_{2} + \underbrace{\frac{dmc^2}{dt_i \varepsilon_H \sqrt{1 - \dfrac{v^2}{c^2}}}}_{3}$$

Expressions 2 and 3 in equation 6I6 are of *indeterminate* form, as with reference to 6I3 and 6I4 are giving $0/0$. Therefore, 6I3 and 6I4 remain valid during the acceleration as well. [2 and 3 are out of the equation.]

6J1   The work intensity of the acceleration equation, Therefore, includes only 4 and 1 and is:

$$w_{acc} = \frac{dmc^2}{dt_p} \left(\frac{1}{\sqrt{1 - \dfrac{v^2}{c^2}}} - 1 \right)$$

The intensity of the electron process remains unchanged, which means the intensity of the neutron process also remains the same: the *Hydrogen* process remains *Hydrogen* process. The intensity of the proton process, however, becomes increased and this results in more intensive generation of the electron processes, but of the same intensity!

6J2   The intensity of the generation of the electron process, as end product of the proton process is:

$$\varepsilon_p = \frac{1}{dt_p \sqrt{1 - \dfrac{v^2}{c^2}}}$$

against the normal $\dfrac{1}{dt_p}$

The more intensive generation of the electron process results in electron process *blue shift* conflict with increased quantum speed of the *Hydrogen* process within the accelerating channel. This conflict itself is generating heat, but it becomes deepened by the additional conflict with the quantum impact of *gravitation*.

The quantum speed values are different, but dt_i the common time system guarantees the conflict. There is no conflict in normal circumstances, as the *IQ* value of the *Hydrogen* process is low.

The benefit of the acceleration is that for having a conflict at temperature of 200-300 °C within the accelerating channel the *Hydrogen* process (in fact the proton process) does not need to be speeded up close to the *Earth* quantum speed value.

The *Hydrogen* process is in gaseous format above the *Earth* surface, having $c_H = 299{,}792$ km/sec quantum speed, equal to the quantum speed on the *Earth*.

Because of the conflict, the acceleration increases the quantum speed of the elementary process.

For having a quantum speed $c_H = 383,870$ km/sec, equal to the quantum speed of the *pyramid* in Section 5, the conflict shall correspond to

$$n\frac{c_H^2}{\varepsilon_H} = \frac{c_x^2}{\varepsilon_H} = \frac{c_H^2}{\varepsilon_x};$$

with $n = 1.6395$

6J3
Ref.
S.5

The intensity of the electron process of the *Hydrogen* process cannot be changed. It stays *Hydrogen* process and the intensity coefficient remains of infinite high value.
This quantum speed increase, however, would be equivalent to the increase of the intensity coefficient of the *Hydrogen* process by n.

This n-value in 6J3 would correspond to $\varepsilon_x = \dfrac{\varepsilon_H}{n} = \dfrac{1}{n}\dfrac{dt_{Hn}}{dt_{Hp}};$ in this way: $dt_{nx} = \dfrac{1}{n}\cdot dt_{Hn}$

6J4

The time decrease in 6J5 would mean increased neutron process collapse – result of the acceleration.

$$dt_{nx} = \frac{dt_{Hn}}{\sqrt{1-\dfrac{v^2}{c^2}}}$$

6J5

With reference to 6J3, the formula in 6J5 gives the speed of the acceleration: v

The theoretical importance here in 6J5 is the relation of the neutron and the proton processes. Because the supposed intensity of the electron process is increased, the neutron process, which is neutral, becomes driven to collapse by increased intensity. The reason for the increase is the acceleration.

$n = \dfrac{1}{\sqrt{1-\dfrac{v^2}{c^2}}};$	gives speed increase values: by $v = c\sqrt{\dfrac{n^2-1}{n^2}}$	n	v (km/sec)	c_x (km/sec)	
		1.6395	237,574	383,870	6J6
		1.3	191,558	341,815	
		1.1	124,892	314,424	

v in the above table is the speed value of the acceleration, c_x is the quantum speed of the elementary process of the *Hydrogen* process – consequence of the acceleration.

The work intensity need of the acceleration of the *Hydrogen* process, in line with 6J1 is: $w_{acc} = \dfrac{dmc_H^2}{dt_p}(n-1)$

6K1

In absolute terms for easier use it will be in simple: $W_{acc} = \dfrac{dmc_H^2}{dt_p\varepsilon_p}(n-1) = mc_H^2(n-1)$

6K2
Ref.
6J4

The intensity value of the electron process corresponds to the speed value in 6J4 $c_x = c_H\sqrt{n}$;   and $E_H = m\cdot nc_H^2$

6K3

The calculations below demonstrate that the energy potential of the conflicting *Hydrogen* electron process is always higher than the work need of the acceleration.

The absolute values of the proton and the electron processes in Table 6.3 well representing also the intensity relations.

The absolute energy capacity of the electron process is: $\dfrac{dmc_x^2}{dt_i\varepsilon_i\varepsilon_x}\left(1-\sqrt{1-\dfrac{(c_x-i_x)^2}{c_x^2}}\right);$ where $dt_i\varepsilon_i = 1$;   and also   taken: $\left(1-\sqrt{1-\dfrac{(c_x-i_x)^2}{c_x^2}}\right) \cong 1$

6K4

The work benefit of the acceleration is the impact of *gravitation*.
The additional conflict, caused by the *blue shift* impact of *gravitation,* can be taken away by cooling.

n	the speed value of the acceleration v	the resulting quantum speed of the elementary process c_x	the absolute work value of the acceleration W_{acc}	the energy potential of the conflict E_H	$\dfrac{E_H}{W_{acc}}$
1	0	299,792.0	0	8.99E+10	
1.1	124,892.7	314,424.5	8.99E+09	9.89E+10	11.00
1.3	191,558.4	341,815.5	2.70E+10	1.17E+11	4.33
1.5	223,451.8	367,168.7	4.49E+10	1.35E+11	3.00
1.7	242,438.7	390,880.9	5.75E+10	1.47E+11	2.56
1.9	254,909.9	413,234.8	7.19E+10	1.62E+11	2.25
2	259,627.5	423,969.9	8.99E+10	1.80E+11	2.00
4	290,272.4	599,584.0	2.70E+11	3.60E+11	1.33
6	295,598.9	734,337.4	4.49E+11	5.39E+11	1.20
8	297,440.7	847,939.8	6.29E+11	7.19E+11	1.14
10	298,289.3	948,025.5	8.09E+11	8.99E+11	1.11
11	298,550.6	994,297.6	8.99E+11	9.89E+11	1.10
12	298,749.2	1,038,510.0	9.89E+11	1.08E+12	1.09
13	298,903.7	1,080.915.4	1.08E+12	1.17E+12	1.083
13.76	299,000.0	1,112.305.1	1.15E+12	1.24E+12	1.078

Table 6.3

Table 6.3

The intensity of the *gravitation* impact is:

6K5
$$\frac{dmc^2}{dt_i \varepsilon_g}\left(1-\sqrt{1-\frac{(c-i)^2}{c^2}}\right); \quad \text{with intensity coefficient } \varepsilon_g = 1$$

Ref. S.1.7 The increased by the intensity of the generation of the electron process *blue shift* surplus and the resulting conflict increases the speed of the quantum communication within the accelerating channel. In this way, with reference to Section 1.7, the *space-time* of the accelerating channel corresponds to the increased quantum speed of the *Hydrogen* process. This is the reason for the heat generation.

[Otherwise the intensity of the accelerated electron process of the *Hydrogen* would be:

6J6
$$\frac{1}{\sqrt{1-\frac{v^2}{c^2}}}\frac{dmc^2}{dt_i \varepsilon_H}\left(1-\sqrt{1-\frac{(c-i)^2}{c^2}}\right); \quad \text{which is of indeterminate form!]}$$

S. 6.6

6.6

***Hydrogen* acceleration and space-time**

The *inflexion point* as the status of relative rest means $\lim dt_o = 0$ and $\lim ds_o = 0$

6L1
6L2 with speed value of quantum communication $\lim c_o = \infty$.

The status of the natural *Hydrogen* process on the contrary means $\lim dt_H = \infty$ and $\lim ds_H = \infty$ with quantum speed $\lim c_H = 0$

This is the reason *plasma* is the start and *Hydrogen* process is the end of the elementary evolution - with similarities:

Plasma has its *Quantum Membrane* of infinite high

6L3 intensity with $\lim IQ = \dfrac{c_{pl}^2}{\varepsilon_{pl}} = \infty$

And the natural *Hydrogen* process has
its *Quantum Membrane* of infinite low intensity with $\quad \lim IQ = \dfrac{c_H^2}{\varepsilon_H} = 0 \qquad$ **6L4**

Characterising both of *Quantum Membrane* in absolute terms as $\quad \dfrac{dmc_{pl}^2}{dt_{pl}\varepsilon_{pl}} = N \dfrac{dmc_H^2}{dt_H \varepsilon_H}; \ \text{and} \ \lim N = \lim \dfrac{c_{pl}^2 \varepsilon_H}{\varepsilon_{pl} c_H^2} = \infty \qquad$ **6L5**

The *plasma*, status of relative *rest*, is in live connection with the *Hydrogen* process, status of relative *motion:*

Relative rest	Relative motion
is characterised by quantum speed and *IQ* of infinite high values (with intensity also of infinite high value, meaning: $\lim \varepsilon_{pl} = 0$);	is characterised by quantum speed, *IQ* and intensity of infinite low values ($\lim \varepsilon_H = \infty$).

The *Hydrogen* process on the surface of the *Earth* is of $c = 299{,}792$ km/sec:
the result of the quantum impact of *gravitation.*

The electron process of the *Hydrogen* process is in surplus as the intensity of its neutron process is $\lim \varepsilon_{nH} = 0$. *Gravitation* and the *blue shift* surplus of the *Hydrogen* process are in conflict – granting in this way energy to the *Hydrogen* process.

$c = 299{,}792$ km/sec, the natural quantum speed value of the surface of the *Earth*, establishes our natural time and space-time.

The intensity of the natural *Hydrogen* process is:	The intensity of the processes on the *Earth* surface:
$\lim \dfrac{dmc_H^2}{dt_H} = 0$	$\dfrac{dmc^2}{dt}; \quad$ with $c = 299{,}792$ km/sec and $dt = 1$

6L6

With speeding up the *Hydrogen* process in an accelerator up to v
close to $c = 299.792$ km/sec, the internal energy intensity value becomes:

$$e = \frac{dmc^2}{dt\sqrt{1 - \dfrac{v^2}{c^2}}} \quad \text{as the time count of the system in motion is slowing down.}$$

6L7

The speeding up in an accelerator is not about double speeding.

The natural $c = 299{,}792$ km/sec established by *gravitation* is the speed value of the quantum communication on the surface of the *Earth*. Otherwise, the quantum speed of the *Hydrogen* process would be $\lim c_H = 0$.

Acceleration for the count of external energy with reference to 6I2 increases the energy content of the subjects. The increased intensity of the generation of the electron process, the surplus of the *blue shift* impact, is in conflict with the quantum impact of *gravitation*!

Ref. 6I2

Gravitation as natural process driven by *plasma* establishes the intensity balance of the elementary *Hydrogen* process on the *Earth* surface. This balance, modified by acceleration, becomes conflicting with *gravitation*. This conflict in energy intensity terms is more than the work intensity demand to be spent for the speeding up of the *Hydrogen* process.

The established by *gravitation* space-time is changing as the consequence of acceleration. There is no impact and no conflict without acceleration.
The status of the *Hydrogen* process in this case corresponds to the space-time controlled by *gravitation*.

7
Isotopes and *isotope* rehabilitation

Isotopes are elementary processes with damaged elementary balance.

7.1
Beta radiation

Beta radiation and *beta* isotope mean the *electron process* and the *anti-electron process blue shift* drives of the elementary processes are damaged.

Electron process deviation may result in (*minus*) and (*plus*) *beta radiation*:
- with *missing intensity* of the *blue shift* drive, or
- with *surplus*, the generating electron process *blue shift* drive is of intensity more than to be used in the elementary process.

  In both cases the quantum speed of the elementary process is out of norm. In the first case is lower, in the second is higher than the standard value of the elementary process.

The rehabilitation of the elementary balance
- in the case of *blue shift* surplus means releasing the *blue shift* surplus,
- in the case of missing *blue shift* impact it means taking electron process *blue shift* drive and proton process cover from the environment, from other elementary processes.

7.1.1. Beta(-) radiation means the elementary process is with missing electron process *blue shift* impact.

The intensity of the drive of the neutron collapse is not sufficient:

$$e_e = \frac{dmc_x^2}{dt_i \varepsilon_x}\left(1 - \sqrt{1 - \frac{(c_x - i_x)^2}{c^2}}\right)$$

7A1

The elementary process is damaged:

The *IQ* (*Intensity Quotient*) value of the isotope is less than it is necessary for the elementary process:

$$IQ_x = \frac{c_x^2}{\varepsilon_x}; \text{ and } \varepsilon_x = \frac{\varepsilon_p}{\varepsilon_n}\sqrt{1 - \frac{(c - i)^2}{c^2}}$$

7A2

7A3

There is a balance deviation within the proton/neutron and in parallel within the anti-neutron/anti-proton process relations.

The loss on electron process intensity within the elementary process is the reason causing *beta(-)* radiation.

If the elementary process is losing on the intensity of the *blue shift* drive of the electron process, the anti-electron processes, as controlling function will be with increased intensity. With reference to 3B1, in the process/anti-process relations, the relation is reciprocal.

$$\varepsilon_{a-e} = \frac{1}{\varepsilon_e}$$

Ref.
3B1

The self-control of the elementary process tries to rehabilitate the deviation.

The *(d-s-b)* neutron collapse within the proton/neutron process relation remains without sufficient electron process drive. The *(u-c-t)* anti-proton collapse within the anti-neutron/anti-proton process relation is driven by anti-electron process *blue shift* drives of increased intensity for increasing the intensity of the proton process through the intensity increase of the anti-proton collapse.

Explanation:

➢ There is a necessary balance between the expansion and the collapse processes within the proton, neutron, anti-proton and anti-neutron processes. Anti-processes are for elementary control and for keeping the balance.

Elementary expansion has its intensity value, followed by its electron processes *blue shift* drive – in normal and anti-directions as well – a function of the intensity of the expansion.

Collapses are driven. Proton and neutron processes both have a complete expansion-collapse quark cycle plus the proton process has in addition an expanding and the neutron process a collapsing quark process.

On the anti-side the anti-proton process has the additional quark collapse and the anti-neutron process has the additional quark expansion process.

➢ Collapse cannot remain without drive, otherwise it stops.

In normal direction the electron process of the additional *(t-c-u)* expansion of the proton process gives the drive to the additional *(d-s-b)* collapse within the neutron process; in anti-direction the anti-electron process *blue shift* drive of the additional *(b-s-d)* expansion of the anti-neutron process loads the remaining without drive *(u-c-t)* collapse of the anti-proton process.

➢ The intensities of the electron process or the anti-electron process *blue shift* drives establish the internal expansion/collapse relation in each direction – in normal and anti-elementary relations as well,

   ➢ by the relation of the intensity of the *(t-c-u)* proton expansion as drive and the intensity of the *(d-s-b)* neutron collapse as driven; and

   ➢ by the relation of the intensity of the *(b-s-d)* anti-neutron expansion as drive and the intensity of the *(u-c-t)* anti-proton collapse as driven.

These should be the internal electron process and the anti-electron process intensity values since these are the drives, which those certain collapses need.

➢ The intensities of the electron and anti-electron processes are quasi reciprocal, as

<table>
<tr><td>in the normal elementary case:</td><td>in the anti-elementary case:</td></tr>
<tr><td>$$\varepsilon_e = \frac{\varepsilon_{p(t-c-u)}}{\varepsilon_{n(d-s-b)}} \sqrt{1 - \frac{(c_x - i_x)^2}{c_x^2}}$$</td><td>$$\varepsilon_{a-e} = \frac{\varepsilon_{a-n(b-s-d)}}{\varepsilon_{a-p(u-c-t)}} \sqrt{1 - \frac{(c_x - i_x)^2}{c_x^2}}$$</td></tr>
</table>

7B1
7B2

since normal and anti-processes are harmonised, as the very basis of the anti-relations:

7B3
$$\varepsilon_{p(t-c-u)} = \varepsilon_{a-p(u-c-t)}; \quad \text{and} \quad \varepsilon_{a-n(b-s-d)} = \varepsilon_{n(d-s-b)}$$

The explanation again is about the physical fact, that the identical collapse needs identical drive:

7B4
with reference to 7B3: $\varepsilon_{a-p(u-c-t)} = \varepsilon_{p(t-c-u)}$

7B5
7B6
with reference to 7B3: $\varepsilon_{n(d-s-b)} = \varepsilon_{a-n(b-s-d)}$

Ref.
Diag.
7.1
$0 < \lim \varepsilon_e < \infty \quad \text{and} \quad \infty > \lim \varepsilon_{a-e} < 0$ As it is presented in Diag. 7.1.

The entropy and the surplus related consequences of the balance have not been reflected in the above.

Beta(-) means: the *(d-s-b)* collapse of the neutron process will be not sufficiently driven.

This does not disrupt the proton/anti-proton and the neutron/anti-neutron harmony, since while the intensity of the electron process *blue shift* drive becomes less (as being damaged), its reciprocal value, the intensity of the anti-electron *blue shift* impact reflects the change.

The *(t-c-u)* expansions run in parallel with *(u-c-t)* and *(d-s-b)* collapses with *(b-s-d)* in proton/anti-proton and neutron/anti-neutron relations. The *(d-s-b)* collapse of the neutron process will be driven by less intensity than the standard value. The *(u-c-t)* collapse of the anti-proton process however will be driven by increased intensity in order to compensate the loss. The isotope status is result of the damage in the proton/neutron and the anti-neutron/anti-proton relations.

The consequence of the damage depends on the characteristics of the elementary process.

➢ In the case of elementary processes *with proton process dominance* the damage is equivalent to less value of electron process *blue shift* surplus within the elementary process. The intensity of the anti-electron process *blue shift* drive will be increased for self-correction, following the reciprocal rule.

The rehabilitation is easy since the elementary process is with electron process *blue shift* drive surplus. At the same time the provision of the anti-neutron process cover might be more time consuming.

➢ In the case of neutron process dominance the consequences are different:

The *(d-s-b)* collapse of the neutron process remains without sufficient drive, while the neutron process dominancy needs it. The intensity of the anti-electron process drive will be increased for the count of the internal energy of the elementary process.

The dominance in the neutron process generates anti-electron process *blue shift* surplus, which will be step by step rehabilitating the balance. The rehabilitation in this case however is more time consuming and difficult, as the provision of the sufficient anti-neutron process cover intensity needs time.

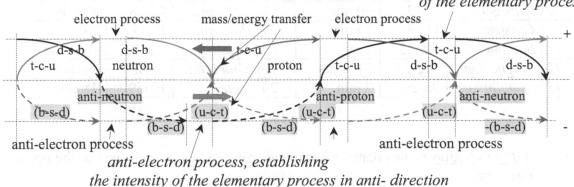

electron process establishing the intensity of the elementary process

anti-electron process, establishing the intensity of the elementary process in anti- direction

Diag.7.1

Diag. 7.1

The symptoms and the reasons of the *beta(-)* radiation in simple terms:
➢ damage in the proton/neutron intensity relation;
➢ reduced drive of the neutron process

As consequence: the elementary process is in *need* in electron process *blues shift* impact and proton process cover!

Beta(-) isotope is looking for *blue shift* impact and proton process cover from external elementary source, as the internal balance is damaged.

Quantum communication with elementary processes with proton process dominance can speed up the rehabilitation process.

Elementary processes with proton process dominance provide electron process *blue shift* drive – from the generated surplus – and proton process cover to the neutron process of the damaged elementary relation = formulating this way in fact the elementary process of the donor element. Taking over this way the damage of the missing electron process *blue shift* drive (*beta-*) they assist to the rehabilitation of the damaged elementary processes.

Isotopes with (*beta-*) means weakened electron process *blue shift* drive.

Neutron processes of the damaged elementary structure have been impacted by external electron process drive and proton process cover. This helps:

(1) to the internal electron process and proton process cover to drive and cover the remaining neutron process demand;

(2) the anti-neutron processes cover and the generating anti-electron process *blue shift* drive this way will be in balance with the anti-proton collapse and proton expansion. (The external electron process *blue shift* support with proton process cover means, the driven neutron process will be representing/belonging to the assisting elementary process. Neutrons are neutral.)

The electron process *blue shift* surplus of the assisting elementary process with proton process dominance speeds up the self-rehabilitation of the donor elementary process.

There are elementary processes with proton process dominance and with close to balanced state, which can provide the missing electron process *blue shift* impact and proton process cover. They will be damaged, but their self-rehabilitation period is short, as it is presented in Table 1 below.

PN		isotope with the longest	*half-life*	PN	Elements	isotope with the longest	*half-life*
1	*Hydrogen*		$\approx 10^{-22}$ sec	14	*Silicon:*	Si-31	2.5 hours
6	*Carbon:*	C-11	21 min	16	*Sulphur:*	S-38	3 hours
7	*Nitrogen:*	N-13	10 min	17	*Chlorine:*	Cl-38	38 min
8	*Oxygen:*	O-15	2 min	19	*Potassium:*	K-43	1 day
11	*Sodium:*	Na-25	1 min	20	*Calcium:*	Ca-45	163 days
12	*Magnesium:*	Mg-28	21 hours	22	*Titanium:*	Ti-45	3 hours
13	*Aluminium:*	Al-29	7 min	26	*Iron:*	Fe-59	45 days

Table 7.1

Table 7.1

S. 7.1.2

7.1.2 *Beta(+)* radiation means damage in the electron process as well, just in the opposite direction.

The damage means increased electron process *blue shift* impact and surplus, result of the intensity increase of the proton process.

Electron process surplus is equivalent to electron process intensity increase:

7C1

$$n\frac{dmc^2}{dt_i \varepsilon_x}\left(1-\sqrt{1-\frac{c_x-i_x)^2}{c_x^2}}\right) = \frac{dmc^2}{dt_i \dfrac{\varepsilon_x}{n}}\left(1-\sqrt{1-\frac{c_x-i_x)^2}{c_x^2}}\right); \quad n>1$$

Less intensity coefficient value results in increasing intensity!

7C2

The definition of the value of the intensity and the *blue shift* conflict of the surplus results in certain intensity increase indeed:

$$n \cdot \varepsilon_e = n\frac{\varepsilon_{p(t-c-u)}}{\varepsilon_{n(d-s-b)}}\sqrt{1-\frac{(c_x-i_x)^2}{c_x^2}}$$

The reciprocal rule works and the self-correction process starts.

With the increase of n as extra surplus, causing $\beta(+)$, the anti-direction reduces the intensity by the increase of n:

The anti-process is controlling the elementary balance and establishing:

$$\varepsilon_{a-p(u-c-t)} = n \cdot \varepsilon_{p(t-c-u)}$$

$$\frac{1}{n}\varepsilon_{a-e} = \frac{\varepsilon_{a-n(b-s-d)}}{n \cdot \varepsilon_{a-p(u-c-t)}}\sqrt{1 - \frac{(c_x - i_x)^2}{c_x^2}}$$

7C3

which tries to rehabilitate the deviation.

The elementary process remains the same therefore the intensity of the electron process should remain the same, just with the unnecessary surplus. Reference here especially should be made to the neutron and the anti-proton processes which are passive, neutral ones as they are driven. Therefore the damage is disappearing.

Isotopes with (*beta+*) are with strong proton process and electron process intensity.

Taking away electron process *blue shift* drive and proton process cover, the generating anti-electron process *blue shift* impact and anti-neutron process cover will initiate the anti-proton process collapse of less intensity (as part is taken away) and contribute even by this to re-establishing the balance. The anti-proton process, driven by less intensity generates proton process of less intensity, source of the increased electron process *blue shift* impact.

Beta(+) as intensity increase is released and might drive neutron processes of other elementary processes as part of quantum communication.

Examples can demonstrate the case:

Elementary process	Isotope from	Isotope to	rad	Stable status	Isotope from	Isotope to	rad	Comm.
Selenium	Se-65	Se-75	$\beta+$	Se-79	Se-79	Se-92	$\beta-$	
Krypton	Kr-69	Kr-79	$\beta+$	Kr-84	Kr-85	Kr-101	$\beta-$	
Strontium	Sr-73	Sr-85	$\beta+$	Sr-88	Sr-89	Sr-104	$\beta-$	
Zirconium	Zr-79	Zr-89	$\beta+$	Zr-92	Zr-93	Zr-108	$\beta-$	
Technetium	Tc-85	Tc-98	$\beta+$	Tc-99	Tc-100	Tc-115	$\beta-$	
Rhutenium	Ru-87	Ru-97	$\beta+$	Ru-101	Ru-105	Ru-118	$\beta-$	
Rhodium	Rh-89	Rh-102	$\beta+$	Rh-103	Rh-104	Rh-121	$\beta-$	
Palladium	Pd-91	Pd-101	$\beta+$	Pd-106	Pd-107	Pd-123	$\beta-$	
Tin	Sn-99	Sn-119	$\beta+$	Sn 120	Sn-121	Sn-137	$\beta-$	
Antimony	Sb-103	Sb-122	$\beta+$	Sb122	Sb-122	Sb139	$\beta-$	
Iodine	I-108	I-1027	$\beta+$	I-127	I-127	I-144	$\beta-$	α ref.
Caesium	Cs-112	Cs-133	$\beta+$	Cs-133	Cs-133	Cs-151	$\beta-$	
Promethium	Pm-128	Pr-144	$\beta+$	Pr-145	Pr-146	Pr-163	$\beta-$	α ref
Samarium	Sm-130	Sm-143	$\beta+$	Sa-150	Sa-151	Sa-165	$\beta-$	α ref
Europium	Eu-132	Eu-152	$\beta+$	Eu-153	Eu-154	Eu-165	$\beta-$	
	Se-79, Kr-85, Sr-90, Zr-93, Tc-98, Ru-106, Rh-106, Pd-107, Sn-121, Sb-125, I-129, Cs-134, Cs-137, Cs-144, Pm-147, Sm-151, Eu-154.Eu-155 are fission product							
Other elementary processes								
Carbon	C-9	C-11	$\beta+$	C-12	C-14	C-22	$\beta-$	
Oxygen	O-13	O-15	$\beta+$	O-16	O-19	O-26	$\beta-$	
Magnesium	Mg-20	Mg-23	$\beta+$	Mg-24	Mg-27	Mg-37	$\beta-$	
Silicon	Si-22	Si-28	$\beta+$	Si-28	Si-31	Si-44	$\beta-$	
Sulphur	S-28	S-32	$\beta+$	S-32	S-35	S-45	$\beta-$	
Chorine	Cl-28	Cl-35	$\beta+$	Cl-35	Cl-35	Cl-51	$\beta-$	
Calcium	Ca-34	Ca-40	$\beta+$	Ca-40	Ca-45	Ca-57	$\beta-$	
Vanadium	Va-44	Va-50	$\beta+$	Va-51	V-52	V-63	$\beta-$	

Table 7.2

Table 7.2

In the case of the rehabilitation of isotopes, the electron process drive, the proton process cover and the anti-process back-up control are the keys for re-establishing the elementary balance of the processes.

S.
7.2

7.2
Alpha radiation

Alpha radiation and alpha isotopes mean increased internal mass/energy balance conflict.

This is not about the deviation of the intensities of the electron and the anti-electron process *blue shift* impacts – to be resolved by cyclical corrections. This kind of damage needs energy/mass intensity release.
The smallest complete elementary cycle is the *Helium* process, as the *Hydrogen* process in normal circumstances does not have *inflexion* point.
Alpha for certain extent is a kind of more serious form of the *beta (plus)* radiation.
This is a proton process/anti-proton process conflict:
The intensity of the proton process cover is not taken in full by the neutron process and the cover demand of the anti-proton process cannot be provided by the anti-neutron process.

The proton process of increased intensity either generates electron process *blue shift* surplus, or *blue shift* impact of increased intensity. Here the neutron process *(d-s-b)* collapse is not taking the increased intensity of the proton *(t-c-u)* process cover. The anti-neutron *(b-s-d)* process is not capable to generate anti-electron process drive and cover – in its controlling function – to the anti-proton *(u-c-t)* collapse, symmetrical to the increased intensity of the proton process.

Alpha radiation is a release of *Helium* process. The *Helium* process is the smallest complete elementary process for release. *Hydrogen* process cannot be released as it is being in elementary cycle for infinite time.

How can an elementary process, part of the elementary evolution be released by another (damaged) elementary process?
The damaged elementary process (as also all elementary processes are) is in quantum communication with the external *quantum system*.
If the elementary balance is in order, the quantum communication is initiated/regulated by the electron processes of the communicating elementary systems. In the case of damage however the quantum communication is about the solution of the conflict!
The *quantum membrane* of the *alpha* isotope – because the reason of the damage is the increased intensity of the proton process – initiates the release of a complete *Helium* process in its natural way – resolving by this way the conflict.
Releasing *Hydrogen* as process would need infinite time as, $\lim dmc^2 = 0$ and the completion of the neutron process is of infinite duration.
The released by the internal conflict *Helium* process is improving the elementary balance.
Releasing the *Helium* process elementary cycle means releasing it in its natural form, with $c_{He} = 299,924$ *km/sec* speed value of quantum communication.

7D1

The release is impacting the external *quantum membrane* by the difference of the square of the quantum speeds on the *Earth* surface and the natural *Helium* process:

$$\frac{dm(c^2 - c_{He}^2)}{dt_i \varepsilon_{He}} \left(1 - \sqrt{1 - \frac{(c-i)^2}{c^2}} \right)$$

The speed difference is impacting the surrounding environment, and while the *He* process release is of rather limited volume, the impact is with damage and it might be significant.

[In natural way the *Helium* process is the product of *gravitation*, the sphere symmetrical expanding acceleration of the *Earth*. And as such it is product of the *Lithium* process, with higher speed of quantum communication – 343,815 km/sec.

In the *alpha* radiation case the *Lithium* process step and all others are bypassed.]

There are no flying *Helium* particles in the case of *alpha* radiation.

Alpha radiation is caused by the impact of the mass/energy intensity surplus of the *Helium* processes with proton process dominance. The intensity of the proton process is more than the intensity of the neutron process. The developing electron process *blue shift* impact and surplus is impacting the environment.

Alpha radiation is similar to (*beta+*).

With the release of the complete *Helium* process, the anti-electron process *blue shift* impact will be capable to drive the remaining anti-proton collapse demand of the isotope, which this way will be corresponding to the normal intensity of the proton process.

7.3
Gamma radiation

S.
7.3

Gamma radiation is the electromagnetic impact of the *extra* anti-electron process *blue shift* surplus of the damaged elementary processes.

Ref.
3E3

In the case of elementary fission the formulating elementary processes with reference to 3E3 are with proton/neutron (and anti-neutron/anti-proton) process intensity relations out of balance. Isotopes, listed specially in Table 7.2 are products of nuclear fission in reactors with *Uranium 235* fuel.

Ref
Table
7.2

There are two elementary components of the experienced *gamma* radiation:

Ref
S.
7.1.1

➤ the intensity of the neutron process of the isotope is increased; this needs – with reference to Section 7.1.1 – additional electron process *blue shift* impact and proton process cover from other elementary sources, causing *beta(-)* radiation;

➤ the proton process of the isotope however remains of the same intensity; this results in additional anti-electron process *blue shift* surplus (as the neutron process, referring to the previous component, is of increased intensity) reason of *gamma* radiation.

$\varepsilon_{iso} = \dfrac{\varepsilon_p}{\varepsilon_n}\sqrt{1 - \dfrac{(c_x - i_x)^2}{c_x^2}}$	$\varepsilon_{iso} < \varepsilon_x$ $\varepsilon_{iso-anti} > \varepsilon_x$ reason of	$\varepsilon_{n-iso} > \varepsilon_{xn}$ and $\varepsilon_{p-iso} = \varepsilon_{xp}$ $\varepsilon_{n-iso} = \varepsilon_{n-iso-anti}$ and $\varepsilon_{p-iso-anti} = \varepsilon_{xp}$ surplus	7E1 7E2

Gamma usually is escorting (*beta-*) or it is the heavier form of (*beta-*), as the case with the examples in Table 7.2 shows it.

In the case of *gamma* radiation the proton process is the one, which is establishing the elementary process, meaning: the type of the isotope is established by the proton process. Neutron process needs drive!

The elementary process is out of balance:

➤ the increased intensity need of the neutron process of the *isotope* is to be driven by electron processes from external sources [(*beta-*) part]; the intensities of the proton process and the electron process drive are of normal values;

> ➤ the increased intensity of the neutron process generates anti-electron process *blue shift* surplus (*gamma* part) by the increased intensity of the anti-neutron process;

The anti-neutron process is an expansion. It is equal in its intensity value and simultaneous with the neutron process. The anti-neutron process of the isotope is developing anti-electron process *blue shift* impact with *extra* surplus:

7E3
$$e_e = \frac{dmc^2}{dt_i \varepsilon}\left(1 - \sqrt{1 - \frac{(c-i)^2}{c^2}}\right) = x\frac{dmc^2}{dt_i \varepsilon_x}\left(1 - \sqrt{1 - \frac{(c-i)^2}{c^2}}\right); \text{ where } \varepsilon = \frac{\varepsilon_x}{x}; \text{ and } \varepsilon_x > \varepsilon$$

(Higher ε electron process intensity coefficient means less electron process intensity.)

In this *gamma* case the intensity increase of the neutron process will be expiring through the intensity of the anti-neutron process – generating *extra* anti-electron process *blue shift* impact surplus.

The proton process of the isotope, the basis of the formulating new elementary proton/neutron process elementary relation cannot fully cover and the electron process cannot fully drive the neutron process with its increased intensity need. Therefore the intensity of the neutron process shall be changed – decreased, the following way:

> The intensity of the anti-proton process is the one establishing the intensity of the anti-electron process, since the first is the driven and the anti-electron process is the driving tool. As the anti-proton and the proton processes run in parallel (controlling each other) this intensity value corresponds to the intensity of the proton process, the basis for the new elementary process. (The entropy generation for simplicity is not taken into account!)

Ref.
7B3
> With reference to 7B3 $\varepsilon_{p(t-c-u)} = \varepsilon_{a-p(u-c-t)}$; and $\varepsilon_{a-n(b-s-d)} = \varepsilon_{n(d-s-b)}$, the intensities of the anti-processes are equal to the intensities of the normal direction = as basis for the controlling function.

Gamma is the release of the increased intensity through the anti-electron process drive and the anti-neutron process cover!

> The intensity of the proton process does not correspond to the increased intensity demand of the neutron process of the isotope, result of the damage – as the components of the electron process *blue shift* drive are:

7E4
$$e_p = \frac{dmc^2}{dt_p}\left(1 - \sqrt{1 - \frac{i^2}{c^2}}\right); \text{ and } e_n = \frac{dmc^2}{dt_n}\sqrt{1 - \frac{(c-i)^2}{c^2}}\left(1 - \sqrt{1 - \frac{i^2}{c^2}}\right); \quad \varepsilon_x \neq \frac{\varepsilon_p}{\varepsilon_n}$$

[To a certain extent *gamma* is the opposite to *alpha*, where the proton process is of increased intensity]

> The intensity coefficient of the electron process of the isotope is different than the intensity represented by the proton process, the basis of the stabilising elementary relation. The intensity of the neutron process is higher, but in absolute terms the mass/energy exchange of the rehabilitation is still guaranteed: $\Delta E_p = \Delta E_n$

7E5

7E6
$$\frac{dmc^2}{dt_p \varepsilon_p}\left(1 - \sqrt{1 - \frac{i^2}{c^2}}\right) = \frac{dmc^2}{dt_{n+}\varepsilon_{n+}}\sqrt{1 - \frac{(c-i)^2}{c^2}}\left(1 - \sqrt{1 - \frac{i^2}{c^2}}\right)$$

The electron process cannot fully drive and the proton process cannot fully cover the neutron process.

7E6 means: The anti-proton process as collapse is taking the intensity of the anti-electron process *blue shift* impact, generated by the anti-neutron process in line with 7E4.

The anti-neutron process as expansion is generating the anti-electron process.

The covering energy/mass intensity is in line with the mass/energy intensity reserve of the neutron process, with increased need relative to the standard elementary relation.

The neutron process of the isotope is driven by the intensity of the electron process *blue shift* impact and covered by the intensity of the proton process, but only partially. The neutron process is still is in electron process *blue shift* demand and proton process cover need (the reason of the escorting *beta-* impact of *gamma* radiation)

The anti-neutron process is releasing the neutron process intensity surplus through the *extra* surplus of the anti-electron process *blue shift* impact. As the controlling function of the anti-neutron process this release is leading the neutron process to reduced intensity: to reaching the relation of the intensities of the anti-neutron and anti-proton processes [by the standard intensity drive of the anti-electron process] for having the stable intensity of the anti-proton (and proton) processes – re-establishing by that the elementary standard.

The generating but released *extra* anti-electron process *blue shift* surplus is impacting the external *quantum membrane* and generating the experienced *electromagnetic* effect. The frequency of the impact represents the intensity of the anti-electron process *blue shift* surplus. The self-rehabilitation process continues until the intensity of the anti-electron process *blue shift* impact corresponds to the standard intensity need of the anti-proton process.

It is important to note that neutron process dominant elementary processes always have anti-electron process *blue shift* surplus, as the intensity of the proton process is of less value. In this particular isotope case however the generating *extra* surplus is the one causing the additional *quantum membrane* impact.

The "normal" anti-electron process *blue shift* surplus with reference to Section 3.3 formulates the energy intensity background (in other words, the internal energy intensity and *gravity* contribution) of the elementary process; while the electron process *blue shift* surplus of proton process dominant elementary processes is the key for elementary communication.

Ref.
S.3.3

Gamma radiation tries to re-establish the normal elementary process without external support. For this reason, it is impacting the external *quantum membrane* – by the *blue shift* impact of the generated anti-electron process of certain frequency (with decreasing improving tendency), for reaching the normal balanced relation of the intensities of the neutron/anti-neutron and proton/anti-proton processes:

$$\varepsilon_{a-e} = \frac{\varepsilon_{a-n}}{\varepsilon_{a-p}} = \frac{\varepsilon_n}{\varepsilon_p}$$

the neutron process and the anti-neutron process intensities are in the nominator here, as the driving impact in the anti-process comes from the anti-neutron process, which, in intensity terms, represents the neutron process of the isotope stage.

7E7

The frequency value of the electromagnetic impact is: $f = \dfrac{1}{dt_i \varepsilon_{a-e}}$

7E8

Gamma usually is escorting the dominant *beta* radiation.

The examples in Table 7.2 are fission products: *Krypton-85, Zircomium-95, Rhutenium-106, Rhodium-106, Antimomy-125, Iodin-131, Caesium-134,-137,-144, Europium-154,-155 plus Cerium-141,-144, Niobium-95.*

Ref.
Table
7.2

Elementary fission with certain proton process and neutron process intensities cannot produce elementary processes (fission products) with exactly corresponding to balanced proton and neutron process intensity statuses. [Fission products with damaged (unstable) elementary process (structure) therefore usually produce additional isotopes.]

(The reason of the isotopes is the difference between the gradients of the intensities of the proton and the neutron processes of the elementary evolution from the *plasma* to *Hydrogen* process.) As the examples in Table 7.3 below clearly demonstrate, the neutron processes of the isotopes of the fission products are all of increased intensity:

With standard proton process intensities (weights), the summarised intensities of the proton expansion and the neutron collapse (in fact the neutron collapse and the anti-proton collapse) the neutron process intensities of the isotopes are more than that of the standard elementary process.	Se-79 against 78.96 Kr-85 against 83.8 Sr-90 against 87.62 Zr-93 against 91.22 Tc-99 against 98.96 Ru-106 against 101.07 Rh-106 against 102.91 Sn-121 against 118.69	Sb-125 against 121.75 I-129 against 126.90 Cs-134 and Cs-137 and Cs-144 against 132.91 Pm-147 against 145.00 Sm-151 against 150.4 Eu-154 and Eu-155 against 151.96
	The only exception is Pa-107, where the standard intensity is 107.4.	

Table 7.3

Table 7.3

<u>Summarising the consequences of elementary damages,</u>
elementary processes with damaged intensities have three ways for survive:
> ➢ releasing *Helium* process with proton process dominance and releasing electron process *blue shift* surplus for taking off energy/mass capacity as *alpha* and *beta+*,
> ➢ taking electron process *blue shift* drive and proton cover from external elementary processes as *beta(-)*, and
> ➢ impacting the environment, "fighting" by the use of the accumulating anti-electron process *blue shift* energy/mass intensity capacities, as *gamma*.

The distinguishing difference between *gamma* and *beta(-)* is that
> ➢ at *beta(-)*, the rehabilitation takes electron process *blue shift* impact and proton process cover from aside "only" – without the anti-electron process *blue shift* quantum impact to the environment of the damaged elementary process;
> ➢ at *gamma*, the rehabilitation involves anti-process components: elementary process from aside are driving the neutron process of the isotope but in parallel as part of the rehabilitation, the anti-electron process extra *blue shift* surplus of the damaged elementary process is impacting the environment - while the internal balance is improving step by step.

The rehabilitation of isotopes is based on the passive character of the neutron process. (Passive is the anti-proton process as well, but the anti-direction is on the controlling side.)
Isotope rehabilitation is elementary process in elementary cycles.
The balance of elementary cycles is with damage. Cycles are formulating in accordance with the internal and external electron process *blue shift* impacts.

Weights of proton and neutron processes are the indicators of their intensities.
Elementary process cycles continue to happen in both cases: balanced or imbalanced way.
External support helps to drive the neutron processes not sufficiently driven by its internal electron process *blue shift*.
As the neutron process is passive and part of the measured *weight* value of the "element" (exceptionally named this way), any external *blue shift* drive and proton process cover helps to reach the elementary balance, since the properly driven and covered neutron processes do not initiate the external elementary environment for driving and covering.

Elementary imbalance does not stop the elementary process the damage however has its consequence: variety of quantum impacts.

From the point of view of the rehabilitation, *beta(-)* and *gamma* isotopes are similar. But external electron process *blue shift* drive and proton process cover (with reference to Table 7.1) does not stop the *gamma* quantum impact. Elements with proton process dominance move into quantum communication with *gamma* isotopes and speed up the rehabilitation.

7.4

X-ray

X-ray radiation means limited electron process *blue shift* impact to the external *environment*: Bombarding certain elementary processes, like *Fe, Ni, Co, Cu, Zr, Mo,* by electron process *blue shift* quantum impact results in *X-ray* radiation.

In fact this is *blue shift* impact to the environment, generated by the *blue shift* impact of elementary processes. The *blue shift* impact of the elementary processes is result of the externally generated electron process *blue shift* conflict.

ε_x, the intensity coefficient of the electron process of the elementary processes listed above is slightly below *1*. All these are with natural neutron process intensity dominance.

Bombarding them by electron process *blue shift* impact is an additional massive impact on their surface regions. This external impact results in electron process *blue shift* conflict with their acting natural elementary *blue shift*: electron process is in surplus, but the neutron process cannot be driven, not just because there is no proton process cover available, but because the elementary processes of these elements do not need additional drive. *X-ray* radiation is an electron process *blue shift* impact of limited intensity.

The frequency of the *blue shifted blue shift* impact depends on the intensity of the electron process of the bombarded elements.

$$e_{X-ray} = \frac{dmc_x^2}{dt_i\varepsilon_x}\left(1 - \sqrt{1 - \frac{(c_x - i_x)^2}{c^2}}\right)^x$$

7F1

The capacity of the impact depends on *x*, the exponent power of the equation.

7.5.

Neutron radiation

Uranium-235 in the nature is with electron process *blue shift* conflict, created by the standard intensity value of the proton process (92) and the reduced intensity of the neutron process (from 146 to 143). In the case of additional *blue shift* impact from elementary processes with electron process *blue shift* surplus, like water or graphite the generating conflict results in the destruction of internal balance: the fission of the *Uranium* elementary process.

Neutron radiation is the external quantum impact of the fission:

The formulating elementary processes, consequences of the fission are in quantum communication with the environment: either missing electron process drive and impacting the environment with *alpha* or *gamma* or being in electron process surplus.

Neutron radiation is not about flying neutrons.

Neutron radiation means taking electron process *blue shift* impact and proton process cover from other elementary processes of the external environment; releasing anti-electron process *blue shift* impact from the generating surplus and causing this way electromagnetic impact.

This is the reason of the heavy damage caused by the "neutron radiation".

As fission is on, the demand in electron process drive (and in parallel) proton process cover from others is also constantly on. Elementary processes, damaged by the fission have *beta*, *alpha* and *gamma* symptoms.

Neutron, gamma and *beta* radiation are in fact of the same character: missing electron process *blue shift* drive and missing proton process cover.

The difference between them is within their appearance:

> *Neutron* radiation happens at elementary process level. The damaged elementary process is missing its neutron process drive and cover.
> *Beta* and *gamma* radiations are at inter-elementary process stage: the developing isotopes, consequence of *neutron* radiation are looking for electron process *blue shift* drive and proton process cover.

S.
7.6

7.6

Rehabilitation of isotopes by external support

As the description about the types of radiation proves, the objective of the rehabilitation process of isotopes is to provide electron process *blue shift* impact and proton process cover to the isotopes.

7G1

$$\frac{dm}{dt_i}(IQ)_1\left[1-\sqrt{1-\frac{(c_1-i_1)^2}{c_1^2}}\right]+\frac{dm}{dt_i}(IQ)_2\left[1-\sqrt{1-\frac{(c_2-i_2)^2}{c_2^2}}\right]+\ldots$$

$$\ldots+\frac{dm}{dt_i}(IQ)_n\left[1-\sqrt{1-\frac{(c_n-i_n)^2}{c_n^2}}\right]=\frac{dm}{dt_i}\left(\frac{c_1^2}{\varepsilon_1}+\frac{c_2^2}{\varepsilon_2}+\ldots+\frac{c_n^2}{\varepsilon_n}\right)\left(1-\sqrt{1-\frac{(c-i)^2}{c^2}}\right);$$

$$\text{as}\left[1-\sqrt{1-\frac{(c_1-i_1)^2}{c_1^2}}\right]\cong\left[1-\sqrt{1-\frac{(c_2-i_2)^2}{c_2^2}}\right]\cong\ldots\cong\left[1-\sqrt{1-\frac{(c_n-i_n)^2}{c_n^2}}\right]\cong\left[1-\sqrt{1-\frac{(c-i)^2}{c^2}}\right]$$

The quantum communication of isotopes with the *Hydrogen, Oxygen, Nitrogen* processes, with the minerals of the *Carbon, Calcium, Sulphur, Silicon* processes improves the elementary balance and this way speeds up the rehabilitation.

These elementary processes with electron process *blue shift* surplus and proton process dominance provide all those elementary quantum impacts, what the damaged elementary processes are missing.

The quantum communication is initiated on the common time system of the electron processes by the electron process *blue shift* demand of the isotopes.

The elementary processes listed above fully support the rehabilitation. They do not even need to be physically mixed with the isotopes. The right distance itself initiates the quantum communication. (The distance is important since the communicating *blue shift* impacts lose on their intensity.)

The most important targets for rehabilitation are the *(-)beta* and *gamma* isotopes.

> Isotopes with *(-)beta* and *gamma* radiation are missing own electron process *blue shift* drive and need external assistance.
> The rehabilitation of the isotopes with (+)*beta* and *alpha* is easier, as the release of the increased intensity helps to reach the balance status in a shorter period of time.

Elementary processes or minerals with electron process *blue shift* surplus have the capacity to support the re-establishing of the elementary balance.

The mix of minerals with rehabilitation effect (with decontamination impact) shall be composed from *Silicon, Calcium, Carbon, Sulphur* elementary processes, from *Oxygen Nitrogen* and *Hydrogen* processes. The mix of these minerals speeds up the rehabilitation process and shortens the half-life of isotopes:

> ➤ provides the electron process *blue shift* impact to the damaged elementary process;
> ➤ and with reference to Section 3.3 and Diag.3.5 the driven (*d-s-b*) quark chain of the neutron processes of the isotopes is covered by the (*t-c-u*) quark process of the proton process of the elementary mix;
> ➤ while the process is controlled by the anti-process of the isotope.

Ref. S.3.3 Diag. 3.5

The elementary processes of the rehabilitation *mix* are in active elementary communication with the isotope. Therefore the process in conventional terms is similar to *absorption*: The electron process *blue shift* surplus of the elementary processes of the minerals of the supporting mix helps to drive and cover not just the natural intensity deficit of the electron process, but also the missing (caused by the damage) *blue shift* impact of the isotope.

There are here however *two* important points conventional considerations are missing:

> ➤ Only the listed above *seven* plus the *Helium* elementary processes are capable to reduce the activity level. The effective way is to have a mix with all these elementary processes acting together as components of the same mineral mix.
> The ideal elementary composition of such a mineral mix for isotope rehabilitation is:
> *Natural Calcium Oxide - CaO*; *Calcium Carbonate - CaCO₃*
> *Kaolinite - $Al_2Si_2O_5(OH)_4$*;
> *Quartz (Silicon dioxide) - SiO_2*; *Sand*
> *Carbon* minerals and *Graphite*;
> All these minerals are with electron process *blue shift* surplus and with intensity coefficient of $\varepsilon_e > 1$.
> The difference between the minerals is in their quantum speed and *IQ* values.
> The *IQ* value of the *Oxygen* process is 8.847E+10, the quantum speed is $c_O = 299,711$ km/sec. The *Hydrogen* process has infinite low *IQ* value and $\lim c_H = 0$ while the *Carbon* process is with $IQ_C = 8.881E+10$; *Calcium* process is with $IQ_{Ca} = 8.921E+10$; and the *IQ* of the *Silicon* process is 8.959E+10.
> The increased *IQ* value of the *Carbon, Calcium* and *Silicon* processes drive the neutrons of the *Oxygen* and the *Hydrogen* processes. The formulating elementary processes will be with electron process *blue shift* surplus. The higher quantum speed value of the mix, higher than the quantum speed on the *Earth* surface results in solid powder format structure even in the case of electron process *blue shift* surplus.
> As result of the support the balance of the elementary processes of the *mix* for the *decontamination* obviously will be damaged. The half-life of the generating "new" isotopes of the mix however is about minutes. (The only exemption is the *Calcium* process, where the half-life of the Ca-45 isotope is 163 days.)
> ➤ The other factor is the quantum communication.
> [Elementary processes communicate with reference to Section 3.3. and Diag.3.5 on their natural way (including obviously the quantum effect as well).]
> Quantum communication however exists and the decontamination effect has its supporting impact without the physical mixing of the elementary processes as well.

Ref. S.3.3 Diag. 3.5

The quantum communication is as specific advantage – it works even from distance. The improving effect in this case depends on the volumes and on the acting distance. And this option extends the so important rehabilitation of the generating by the nuclear industry isotopes.

S.   7.6.1. <u>For proving the statement about the rehabilitation isotopes</u> a decontamination mix
7.6.1   was prepared and presented in details within Annex 7.1.

The experiment was made for the rehabilitation of the *Potassium-40* isotope within *KOH* base, the only mixture with an isotope free for sale.

The rehabilitation of the experiment has two formats:
  ➤ mixing the decontamination mix (*Decomix)* with the *KOH* – for measuring the direct impact; and
  ➤ placing 1 kg container with *KOH* within into a 5 kg *Decomix* bath;

Ref.   the *KOH* within the container was left to quantum communication with the *Decomix*
Annex   only – for measuring the quantum impact.
7.1   The experiment is presented in details within Annex 7.1.

<u>Experiment 1</u>:
(represented by Diag.7.2)
The mixing was made in two ways: in portions 1/1 and 1/2.
The 1/2 relation means the portion of the *Decomix* was half of the *KOH* base.
The initial activity of the KOH was *40.21* cpm.

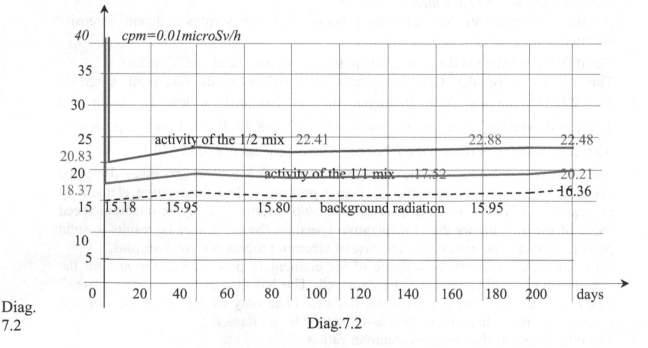

Diag.
7.2   Diag.7.2

The mixing in both cases has resulted in immediate and aggressive heat development.
The activity of the mix was measured, once the heat generation of the mix was over. It has taken a couple of minutes.

The 1/1 mixture had higher temperature increase.

Heat generation means *blue shift* conflict initiated by the *Decomix*.

The increased temperature went down and in parallel the *KOH* has lost its activity.

There was an immediate activity reduction

to *18.37 cpm* in the 1/1 mixture and to *20.83 cpm* in the 1/2 mixture.

The background radiation was *15.18 cpm*.

The *KOH* mixed with the *Decomix* resulted in immediate activity reduction and further reduction for 132 days. For the last 50 days the measured activity was slightly increased by 0.02 $\mu Sv/h$.

This change for such a low activity value as it is for the *KOH* shall be within the tolerance of the measurements even in the case if around 100 measurements were made for a single data. As the measured values can easily be influenced by external factors.

The conclusion is clear: the 1.3 billion year half-life of the *Potassium-40* (*K-40*) isotope was reduced to a couple of minutes!

The other experiment was with the activity measurement of the *KOH* in *Decomix* bath, without mixing it with the *Decomix* powder for the period of 180 days.

Experiment 2:

(represented by Diag.7.3)

The activity of the *KOH* within a separate small 1 kg container was placed into the *Decomix* bath and left to quantum communication only, without mixing shows decreasing activity results as well. The gradient of the reduction of the *KOH* activity is significant, as the diagram below proves it.

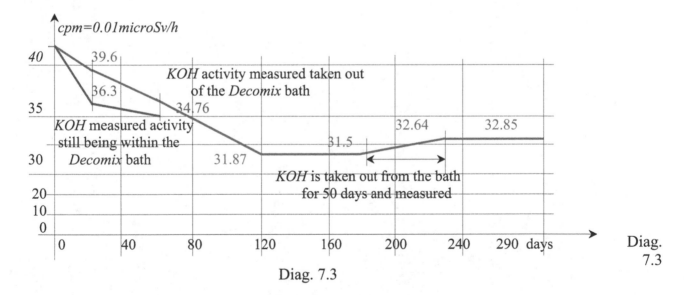

Diag. 7.3

The *KOH* was taken out from the *Decomix* bath for 50 days. The results show a slight increase of the KOH by 0.01 $\mu Sv/h$.

The *1 cpm* measurement unit activity corresponds to 0.01 $\mu Sv/h$.

As the measurements at this activity level are very sensitive to any external impacts, mainly from the background, each data on the diagrams and within the tables of the annex represent the average value of a 80-120 measurement cycle for both experiments.

Ref The description of the use of *Decontamination mix* proposed for the rehabilitation of
Annex *Caesium* isotopes is given in details in Annex 7.2.
7.2 The example of the calculation of the rehabilitation of RBMK graphite stacks by
Annex *Decontamination mix* is given in Annex 7.3.
7.3

8
Quantum impacts of *pyramids* - the power

The quantum communication of the *pyramid* depends on the relations of the territories of the casing and the basic surfaces.

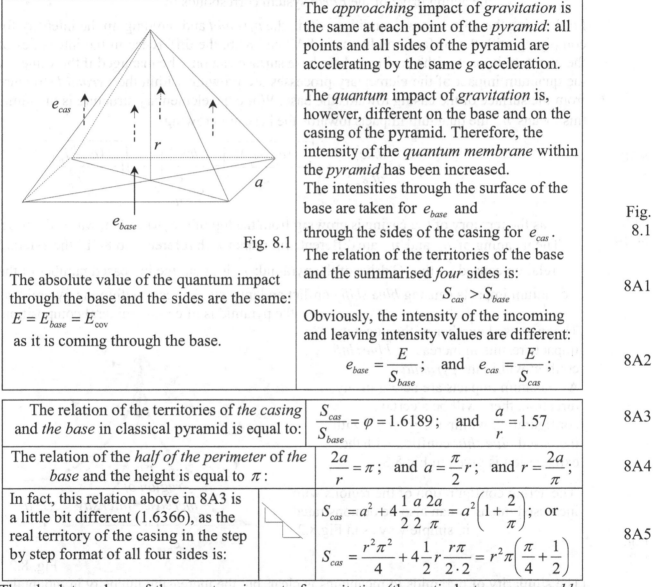

Fig. 8.1

The *approaching* impact of *gravitation* is the same at each point of the *pyramid*: all points and all sides of the pyramid are accelerating by the same g acceleration.

The *quantum* impact of *gravitation* is, however, different on the base and on the casing of the pyramid. Therefore, the intensity of the *quantum membrane* within the *pyramid* has been increased.
The intensities through the surface of the base are taken for e_{base} and through the sides of the casing for e_{cas}.

Fig. 8.1

The relation of the territories of the base and the summarised *four* sides is:

$$S_{cas} > S_{base}$$

The absolute value of the quantum impact through the base and the sides are the same:
$$E = E_{base} = E_{cov}$$
as it is coming through the base.

Obviously, the intensity of the incoming and leaving intensity values are different:

$$e_{base} = \frac{E}{S_{base}}; \quad \text{and} \quad e_{cas} = \frac{E}{S_{cas}};$$

The relation of the territories of *the casing* and *the base* in classical pyramid is equal to:	$\dfrac{S_{cas}}{S_{base}} = \varphi = 1.6189$; and $\dfrac{a}{r} = 1.57$	8A3
The relation of the *half of the perimeter* of the *base* and the height is equal to π:	$\dfrac{2a}{r} = \pi$; and $a = \dfrac{\pi}{2} r$; and $r = \dfrac{2a}{\pi}$;	8A4
In fact, this relation above in 8A3 is a little bit different (1.6366), as the real territory of the casing in the step by step format of all four sides is:	$S_{cas} = a^2 + 4\dfrac{1}{2}\dfrac{a}{2}\dfrac{2a}{\pi} = a^2\left(1 + \dfrac{2}{\pi}\right)$; or $S_{cas} = \dfrac{r^2\pi^2}{4} + 4\dfrac{1}{2}r\dfrac{r\pi}{2\cdot 2} = r^2\pi\left(\dfrac{\pi}{4} + \dfrac{1}{2}\right)$	8A5

The absolute values of the quantum impact of *gravitation* (the anti-electron process *blue shift* impact of the elementary processes of the *Earth*) through the base and the casing are equal, but their surface related intensities depend on the relation of the territories of the surfaces. The intensity of the quantum impact within the *pyramid* is building up starting from the casing. It cannot be differently, since the casing is communicating with the external quantum system, quantum speed of c_{Earth}. This is the only way for guaranteeing the continuity of the fluent quantum communication.

The building of the *pyramid* is homogenous; it is one and the same elementary process, *granite* or *limestone*.

Ref.
8C14
8A6

The quantum impact through the base of the pyramid is:

$$\frac{dmc_p^2}{dt_i \varepsilon_p}\left(1 - \sqrt{1 - \frac{(c_p - i_p)^2}{c_p^2}}\right) = n_E \frac{dmc_E^2}{dt_i \varepsilon_E}\left(1 - \sqrt{1 - \frac{(c_E - i_E)^2}{c_E^2}}\right);$$

8A7

The $c_p \cdot \varepsilon_p = c_E \cdot \varepsilon_E = const$ conditions for the *pyramid* and for the external quantum system are met;
the acting drive of the *Earth* system corresponds to:

$$IQ_p = n_E \cdot IQ_E = \frac{c_E^2}{\dfrac{\varepsilon_E}{n_E}}$$

For keeping the $c \cdot \varepsilon = const$ condition within the *pyramid* and building up the intensity for complying with the surface conditions in 8A2 and 8A6, the difference in the intensities of the quantum impact of the casing and the base surface can only be managed if the values of the quantum impact of the elementary processes are growing within the *pyramid* structure from the surface of the casing towards the base. While the elementary structure is the same, this means n_p the quantum impacts towards the base are growing:

8A8

$$n_p \frac{dmc_p^2}{dt_i \varepsilon_p}\left(1 - \sqrt{1 - \frac{(c_p - i_p)^2}{c_p^2}}\right) = \frac{dmc_p^2}{dt_i \dfrac{\varepsilon_p}{n_p}}\left(1 - \sqrt{1 - \frac{(c_p - i_p)^2}{c_p^2}}\right);$$

Ref.
8C14

as the territory of the casing is growing from the top of the *pyramid* towards the base. The meaning of n_E and n_p are different: n_E marks with reference to 8C14 the external relation caused by the *IQ* drive of the pyramid; n_p is about the increased number of the quantum impacts causing *blue shift* conflict and intensity increase within the pyramid, as the pyramid is of certain mineral composition.

The increased number of the quantum impacts results in increasing *blue shift* conflict within the *pyramid*.
As quantum impacts are acting in any direction, there will be a certain configuration within the *pyramid* with increased *blue shift* conflict, with the explanation is given in Fig.8.3.

The acting configuration of the regions with increased intensity can also be demonstrated in simple way as in Fig.8.2:

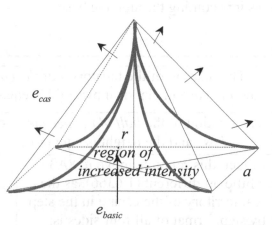

Fig.
8.2

Fig. 8.2

[The similarity of pyramids to mountains is clear, but the increasing intensity in mountains is the result of the composition of elementary processes with increasing intensity.]
A *Pyramid* is a kind of "quantum pump", pumping out the quantum impact of *gravitation*. This means that, in order to ensure the continuity of the quantum impact of the quantum membrane and the communication with the external quantum system through the casing, the feeding by *gravitation* through the basic surface of the *pyramid* shall be increased. The intensity of the quantum impact through the base is the same as in other normal places, but with increased number.

With reference to Section 4.4 and the explanation in Exp.4.1, *granite,* the building stone of the pyramid, is a composition of elementary processes with natural x (the relation of quantum speed values) and n (volumes) establishing a close to balance elementary structure.

For an elementary composition with two elementary processes the values of x and n are:

$$n = \frac{\varepsilon_A}{\varepsilon_B}; \quad \text{as } \varepsilon_A = \frac{1}{\varepsilon_{a-A}}; \text{ and } \varepsilon_B = \frac{1}{\varepsilon_{a-B}}; \quad n = \frac{\varepsilon_{a-B}}{\varepsilon_{a-A}} \quad \bigg| \quad x = \frac{c_B^2}{c_A^2} \qquad \text{8B1}$$

Granite is unique as the structure is with many of the elementary processes. The generated space-time belongs to the composition of the elementary mix of the minerals of the pyramid.

The pyramid has been built up from *granite* minerals with general formula of *Feldspars* with the main component of: $KAlSi_3O_8 -$ $NaAlSi_3O_8 -$ $CaAl_2Si_2O_8$	SiO_2 - 72.4 % Al_2O_3 - 14.42 % K_2O - 4.12 % Na_2O - 3.69 % CaO - 1.82 % FeO - 1.68 % Fe_2O_3 - 1.22 % MgO - 0.71 % TiO_2 - 0.30 % P_2O_5 - 0.12 % MnO - 0.05 %

Element	ε_x	c_x	$IQ = \dfrac{c_x^2}{\varepsilon_x}$
Oxygen	1.01533	*299711**	8.847E+10
Sodium	0.93084	313018	1.053E+11
Magnesium	0.98990	303536	9.307E+10
Aluminium	0.94697	310340	1.017E+11
Silicon	1.00898	*300652*	8.959E+10
Phosphorus	0.95283	309348	1.004E+11
Potassium	0.95932	308335	9.910E+10
Calcium	1.01112	*300334*	8.921E+10
Titanium	0.86133	325403	1.229E+11
Manganese	0.84667	310340	1.138E+11
Iron	0.88347	321300	1.169E+11

Table 8.1

Table
8.1

All elements in Table 8.1 except *Ti, Mn* and *Fe* are with proton and neutron process intensities around equilibrium status, with quantum speed and intensity values close to each other. The components of the minerals communicate and give a composition of solid, almost unbreakable granite structure.

The integrated intensity of the quantum membrane within the granite *pyramid* can be calculated as the sum of the proportions of the components and the subcomponents of the mix is:

$$(n_1 + n_2 + ... + n_n) = 1 \quad \text{which gives}$$

$$0.724 \cdot IQ_{SiO2} + 0.1442 \cdot IQ_{Al2O3} + ... + 0.005 \cdot IQ_{MnO} = IQ_{mix} \qquad \text{8B2}$$

The sum of the proportion of the quantum drive of each of the components gives the unit *IQ* drive of the granite:

$$n_1 \frac{dmc_{x1}^2}{dt_i \varepsilon_{x1}} \left(1 - \sqrt{1 - \frac{(c_{x1} - i_{x1})^2}{c_{x1}^2}}\right) + n_2 \frac{dmc_{x2}^2}{dt_i \varepsilon_{x2}} \left(1 - \sqrt{1 - \frac{(c_{x2} - i_{x2})^2}{c_{x2}^2}}\right) + ... +$$

$$n_n \frac{dmc_{xn}^2}{dt_i \varepsilon_{xn}} \left(1 - \sqrt{1 - \frac{(c_{xn} - i_{xn})^2}{c_{xn}^2}}\right) = \frac{dmc_{mix}^2}{dt_i \varepsilon_{mix}} \left(1 - \sqrt{1 - \frac{(c_{mix} - i_{mix})^2}{c_{mix}^2}}\right)$$

With reference to Table 8.2 the $n = 11$ components of granite give *IQ* drive, value of:

$$IQ_{mix} = IQ_{granite} = 9.0641E + 10 \ km^2/sec^2 \qquad \text{8B4}$$

8B5 For finding the correct intensity values, the intensity coefficients shall be multiplied by the proportions of the components, otherwise the result would be incorrect (as the intensity value would become modified).

$$n_n \frac{c_n^2}{\varepsilon_n} = \frac{\frac{c_n^2}{\varepsilon_n}}{n_n}$$

The multiplication shall not change the intensities of the components.

8B6 $$\frac{n_1 \varepsilon_{x1} + n_2 \varepsilon_{x2} + ... + n_n \varepsilon_{xn}}{n_1 + n_2 + ... + n_n} = \varepsilon_{mix}$$ In the case of the granite structure, the intensity coefficient of the mix is: $\varepsilon_{mix} = \varepsilon_{granite} = 1.00421$

8B7 The coefficient of the intensity of the electron processes and the *IQ* drives of the components are calculated on the basis of the ratio of the subcomponents.

$$\varepsilon_{SiO_2} = \frac{\varepsilon_{Si} + 2 \cdot \varepsilon_O}{3}$$

8B8 The example for the SiO_2 is $IQ_{SiO_2} = \dfrac{IQ_{Si} + 2 \cdot IQ_O}{3}$; and $c_{SiO_2} = \sqrt{IQ_{SiO_2} \cdot \varepsilon_{SiO_2}}$

The speed of quantum communication of the *granite* structure is:

$$c_{granite} = \sqrt{IQ \cdot \varepsilon_{granite}} = 301{,}699.87 \text{ km/sec.}$$

As the calculation proves, granite is a mineral with proton process dominance; with higher speed of quantum communication than the quantum speed on the *Earth* surface. It is similar to the *Si, Ca, Cl, S* processes, just with increased stability of the structure, because of the higher quantum speed and integrated intensity coefficient closer to 1.

The elementary anti-electron process *blue shift* quantum impact of *gravitation* is being effected through the base of the *pyramid*. The quantum speed of the *pyramid*, with reference to 8B8 above, is $c_p = 301{,}699.87$ km/sec; and the integrated intensity coefficient, with reference to 8B5 is $\varepsilon_p = 1.00421$.

8B9 The quantum drive of *gravitation* through the base is $IQ_g = \dfrac{c_c^2}{\varepsilon_g} = \dfrac{299{,}792^2}{1} = 8.9875 + 10 \ km^2/sec^2$

8B10 The impact through the surfaces of the casing, the quantum speed and the intensity of the granite, with reference to Table 8.2 and 8B4 is $IQ_{granite} = 9.0641E + 10 \ km^2/sec^2$.

This is the quantum impact at the external surface of the casing of the *pyramid*.

8C1 $\dfrac{IQ_{gr}}{IQ_g} = \dfrac{9.0641}{8.9875} = 1.00852$ The casing of the *pyramid* is impacting the external quantum system. (If the quotient of the two were less than 1, the pyramid would be the one taking intensity in.)

Therefore, the external *Quantum Membrane* close to the surface areas of the *pyramid* is with generation of slight *blue shift* conflict.

8C2 The quantum impact of the pyramid through the casing to the external *Quantum Membrane* is: $e_{gr} = \dfrac{dmc_{gr}^2}{dt_i \varepsilon_{gr}} \left(1 - \sqrt{1 - \dfrac{(c_{gr} - i_{gr})^2}{c_{gr}^2}} \right);$

The *IQ* quantum drive of *gravitation* with reference to 8C1 is less than the quantum drive of the granite of the *pyramid*. This means that the conflict at the surface of the casing will be generating quantum impacts with the quantum speed of the *Earth* in slightly increased numbers. This increased number here is irrelevant. This is the direct consequence of the quantum communication with difference in the quantum speed and *IQ* values.

8C2 is the intensity impact for the further assessment.

8C2 is also the intensity value, consequence of the quantum load from *gravitation*.

The surface related intensity, loading the pyramid through the base surface is

The surface related intensity leaving the pyramid through the covering surfaces is

$$\frac{dmc_{gr}^2}{dt_i \varepsilon_{gr} f_b}\left(1-\sqrt{1-\frac{(c_{gr}-i_{gr})^2}{c_{gr}^2}}\right) > \frac{dmc_{gr}^2}{dt_i \varepsilon_{gr} f_c}\left(1-\sqrt{1-\frac{(c_{gr}-i_{gr})^2}{c_{gr}^2}}\right); \quad \text{since } f_c > f_b \qquad 8C3$$

The surface related multiplications of the electron process intensities ($\varepsilon \cdot f$) within the equation in 8C3 are virtual values. These are only to prove that at equal numbers and equal electron process *blue shift* intensities of the loading and the leaving impacts, the intensity of the leaving impact of the pyramid is less. For demonstrating the standard electron process *blue shift* impact intensity need of the granite body of the pyramid both sides of the formula in 8C3 are multiplied by the surface value of the casing.

$$\frac{f_c}{f_b}\frac{dmc_{gr}^2}{dt_i \varepsilon_{gr}}\left(1-\sqrt{1-\frac{(c_{gr}-i_{gr})^2}{c_{gr}^2}}\right) > \frac{f_c}{f_c}\frac{dmc_{gr}^2}{dt_i \varepsilon_{gr}}\left(1-\sqrt{1-\frac{(c_{gr}-i_{gr})^2}{c_{gr}^2}}\right); \quad \text{as } f_c > f_b \qquad 8C4$$

This ensures the relation and the impact as per 8C1 and 8C2, but the number of the loading *blue shift* impact is increased. There is a quantum membrane expected to be formulating within the pyramid, since the use is of less intensity and the load is of more.

The intensity and the quantum drive depend on the number of the acting quantum impacts:

$$n\frac{dmc^2}{dt_i \varepsilon}\left(1-\sqrt{1-\frac{(c-i)^2}{c^2}}\right) = \frac{dmc^2}{dt_i \dfrac{\varepsilon}{n}}\left(1-\sqrt{1-\frac{(c-i)^2}{c^2}}\right)$$

Through the increase of the acting quantum drives, the intensity of the generating quantum membrane is also growing.

8C5

With reference to 8C5 above, the quantum load through the basic surface is increased.

8C4 and 8C5 mean that the increase in numbers through the basic surface giving the required impact through the casing shall correspond to the relation of the surfaces

$$\eta = \frac{f_c}{f_b} = \varphi = 1.6189 \qquad 8C6$$

which, in this case, is equal to 8C6, the so-called *Golden Ratio*.

The quantum drive of the *Quantum Membrane* of the pyramid is: $IQ_{pyr} = \eta \cdot IQ_{granite}$

$$IQ_{pyr} = 1.6189 \cdot (9.0641E+10) = 1.46739E+11 \ km^2/sec^2; \qquad 8C7$$

As the granite structure is not changing, the electron process *blue shift* conflict and the increased drive corresponds to the increased speed value of quantum communication, which is equal to: $c_{pyr} = \sqrt{IQ_{pyr} \cdot \varepsilon_{granite}}$ and $c_{pyr} = 383,870$ km/sec, \qquad 8C8

This quantum speed value is higher than that of the *Uranium* elementary process ($c_U = 378,163$ km/sec), the highest among the elementary processes.

(The data for the calculation of the quantum speed above might be taken not precisely enough. Therefore, its value might be slightly different. The approach, however, is certainly correct.)

The surface of the casing of the *pyramid* with $IQ_{granite} = 9.0641E+10 \ km^2/sec^2$ and quantum speed of $c_{granite} = 301,699.87$ km/sec is communicating with the external quantum membrane of the *Earth* $c_{Earth} = 299,792$ km/sec.

There are a couple of points to remember here:

1. Multiplying both sides of the relation in 8C4 by their own corresponding surface value gives equality. This proves the approach and the related calculations are correct, the absolute impacts at both ends are equal.

Ref
8C4

8C9
$$\frac{f_c}{f_b} f_b \frac{dmc_{gr}^2}{dt_i \varepsilon_{gr}} \left(1 - \sqrt{1 - \frac{(c_{gr} - i_{gr})^2}{c_{gr}^2}}\right) = \frac{f_c}{f_c} f_c \frac{dmc_{gr}^2}{dt_i \varepsilon_{gr}} \left(1 - \sqrt{1 - \frac{(c_{gr} - i_{gr})^2}{c_{gr}^2}}\right);$$

2. The driving intensity surplus between the internal membrane of the pyramid and the surface of the casing is:

8C10
$$e_\Delta = \eta \frac{dmc_{gr}^2}{dt_i \varepsilon_{gr}} \left(1 - \sqrt{1 - \frac{(c_{gr} - i_{gr})^2}{c_{gr}^2}}\right) - \frac{dmc_{gr}^2}{dt_i \varepsilon_{gr}} \left(1 - \sqrt{1 - \frac{(c_{gr} - i_{gr})^2}{c_{gr}^2}}\right);$$

With reference to 8C7 and 8B4, the *pyramid* works out the increased intensity of its internal quantum membrane towards the surface of the casing.

8C11
$$\text{from } IQ_{pyr} = 1.46739\text{E}+11 \ km^2/\sec^2 \text{ to } IQ_{gr} = 9.0641E+10 \ km^2/\sec^2$$

This is the potential which ensures the continuity of the quantum communication with the external quantum system (of the *Earth* surface). The acting intensity of the impact is building up from the casing. The number of the quantum impacts of *gravitation* through the base surface depends on the intensity demand of the casing. The *pyramid* is communicating through its casing surface quasi the quantum impact of *gravitation*.

3. The quantum drive intensity surplus is:

8C12
$$\text{In short: } e_\Delta = (\eta - 1)IQ_{gr} = 0.6189 \cdot IQ_{gr} = 5.6097E+10 \ km^2/\sec^2$$

8C13
With the standard quantum speed of the granite taken, the intensity coefficient of the surplus is:
$$\varepsilon_\Delta = \frac{\varepsilon_{gr}}{0.6189} = \frac{1.00421}{0.6189} = 1.6625$$

Pyramids can also be made from other minerals or elementary compositions, like limestone. The principle of the quantum impact is similar, but the speed of quantum communication and the establishing space-time are different, corresponding to the elementary composition of the *pyramid*.

The significance of *granite* as the material of the *pyramid* – alongside with its stability – is the developing by granite increased quantum drive impact.

Limestone, for example, produces quantum impact of less intensity (as the quantum speed of the components – *Ca, C, O* – are of less value) and the stability of the structure is also less.

Diamond, as a building material, could obviously not be used on an industrial scale.

Elementary processes in gaseous state like *Oxygen* and *Nitrogen* or clean elementary processes like *Silicon* or *Calcium* could not be used as they would need a separate frame structure for keeping them in *pyramid* format – which would be interacting with the increased electron process intensity and proton process cover.

How do quantum systems with different quantum speed values communicate?

The main technical points of the *pyramid* structure in Fig 8.1 and 8.2 are:

1. The quantum speed on the *Earth* surface is: $c_{Earth} = 299,792$ km/sec.

   This speed value has been established by the expanding acceleration of the *Earth* (as being in electron process function and communicating anti-electron process impact).

2. The pyramid building is taken by the sphere symmetrical expanding acceleration of the *Earth*. This means that the speed of the all over casing of the pyramid is $i_{Earth} = \lim g\Delta t = c_{Earth}$, with the approaching impact of *gravitation*: g.

   At the same time, the quantum impact of the casing through the all over surface is with: $c_{granite} = 301,699$ km/sec.

The increased relative to the *Earth* surface quantum speed generates certain, slightly increased quantum membrane around the casing of the pyramid.

$$\frac{c_{gr}^2}{\varepsilon_{gr}} = n_E \frac{c^2}{1}; \quad n_E = \frac{1}{\varepsilon_{gr}} \frac{c_{gr}^2}{c^2} = 1.0085 \qquad \text{8C14}$$

3. By the increased number of the impacts of *gravitation*, the developing conflict of the quantum membrane within the pyramid generates quantum speed, corresponding to $c_{pyr} = 383{,}870$ km/sec, as with reference to 8C15 below: $c_{pyr} = c_{gr}\sqrt{1.6189}$

The increased quantum speed value within the pyramid is result of the *blue shift* conflict:

$$\frac{c_{pyr}^2}{\varepsilon_{granite}} = \eta \frac{c_{granite}^2}{\varepsilon_{granite}} \qquad \text{and } \eta \text{ is changing from the value of the } \textit{Golden Ratio} \text{ of the quantum membrane of the } \textit{pyramid} \text{ to the surface with } 1. \qquad \eta = \frac{c_{pyr}^2}{c_{gr}^2} = \varphi = 1.6189 \qquad \text{8C15}$$

The electron process *blue shift* conflict inside the *pyramid* may also generate the flow away of the part of the *blue shift* impact of the surplus. This will not destroy the balance between the *Earth* surface and the quantum system above, since the pyramid, as an "energy pump," will always reload the internal membrane in line with the external demand.

Gravitation is the *blue shift* impact of the surplus of the anti-electron processes of the anti-elementary processes. This impact, in conventional terms, can be called an electromagnetic impact or the background radiation of gravitation. [This is different than the direct *blue shift* impact of the approaching function of the acceleration: The approaching impact is a quasi mechanical effect to the existing quantum system – resulting in *blue shift* impact.]

The quantum impact of *gravitation* is the original anti-electron process *blue shift* surplus of the elementary evolution, which acts through the surface. The anti-electron process with reference to Section 1.2 is losing on the intensity of its elementary drive, but the surplus is constantly being recreated by the direct elementary process. This is the acting drive of the elementary evolution between the *plasma* state and the *Hydrogen* process.   Ref. S.1.2

Having wires or cables with *Iron, Cuprum, Nickel, Zinc* elementary processes within the pyramid, the increased *blue shift* conflict (as the experiment proves) generate electron process *blue shift* impact flow within these cables or wires.   Ref. S.8.3

The *IQ* drive within the pyramid, as consequence of the intensity increase and the *blue shift* conflict, with reference to 8C6, is   Ref. 8C6

$$\frac{c_{pyr}^2}{\varepsilon_{granite}} = \eta \frac{c_{granite}^2}{\varepsilon_{granite}}; \qquad \frac{c_{pyr}^2}{\varepsilon_{granite}} = \eta \frac{c_{granite}^2}{\varepsilon_{granite}} = \eta_{Cu} \frac{c_{Cu}^2}{\varepsilon_{Cu}} \qquad \text{8D1}$$

Taking for example a wire with *Cuprum* process built in within the pyramid, the conflict will be impacting the *Cuprum* process as well. For the balance and the continuity of the quantum communication between the increased quantum impact of the pyramid and the *Cuprum* process, the *IQ* drive of the *Cuprum* process with reference to 8D1 corresponds to: Ref. 8D1

$$1.6187 \frac{301{,}699^2}{1.00421} = \eta_{Cu} \frac{327{,}304^2}{0.85135}; \qquad \text{8D2}$$

$$1.7306\text{E}{+}11 = \eta_{Cu} \cdot (1.2591\text{E}{+}11) \text{ and } \eta_{Cu} = 1.375 \qquad \text{8D3}$$

As $\eta_{Cu} > 1$ there is an electron process *blue shift* surplus generating within the *Cuprum* process wire – proving the propagation of the *blue shift* impact. There is *blue shift* impact potential within the wire inside the pyramid comparing to the external quantum system.

S.
8.1

8.1
IQ and the intensity relations

The internals of the *pyramid* are in conflict; conflicts communicate with increased intensity; increased intensity means high *IQ* value quantum drive and increased speed of quantum communication.

Ref.
S.5

With reference to Section 5, there are numbers of space-times with different quantum speed values within the internal structure of the *pyramid*. The intensity growth is building up from the external surface. The internal regions with increased intensity feed the consumption at the external surface, whatever value the external demand is.

8E1   The *space-time* inside the *pyramid* with increased intensity of $IQ = \dfrac{c^2_{pxx}}{\varepsilon_{pyr}}$ has two options:

➢ either consolidating through the internal structure at the external surface, or
➢ communicating with a space-time of similar intensity.

Consolidating:
The *Earth* surface is with operating $c_{Earth} = 299{,}792$ km/sec quantum speed.

8E2   The conflict: $\dfrac{dmc^2_{pxx}}{dt_i \varepsilon_{pyr}}\left(1 - \sqrt{1 - \dfrac{(c-i)^2}{c^2}}\right)$;   The surface $\dfrac{dmc^2_E}{dt_i \varepsilon_E}\left(1 - \sqrt{1 - \dfrac{(c-i)^2}{c^2}}\right)$

8E3   The gradient of the change is: $\dfrac{d}{df}\dfrac{c^2}{\varepsilon} = \varphi \div 1 = 1.6189 \div 1$

The intensity increase (the temperature) is less and less towards the surface and all intermediate pyramid *space-times* are creating their own quantum membrane. The difference in these quantum membranes is in their quantum speed and intensities. The last one, in fact, is the impact of *gravitation* equal to the intensity and the quantum speed of the space-time above the *Earth* surface.
The internal region is with a kind of *quantum pressure*.
The goal of the consolidation is speed reduction, reaching the quantum speed and the intensity of the *gravitation* beyond the external surface of the pyramid. This, in fact, is not a "classical quantum communication" rather a constant reduction of the quantum speed of the quantum impact itself.
The reduction is a fluent process.

Space-time communication:
Quantum systems are equally within the *pyramid* and outside the pyramid.
There are (infinite) numbers of quantum systems (space-times) with different quantum speed values between. The centre of the *pyramid* has its own quantum system space-time. All these existing space-times with reference to Section 2.3 are impacting the overall quantum system. The electron process *blue shift* impact of the space-times can be utilised by space-times of similar quantum speed and intensity.

Ref.
S.2.3

The consuming of the impact *space-time* can be built up at any "distance".
The transmitted quantum intensity impact of *gravitation* directly communicates with *space-times* of equal intensities and quantum speed values. The higher the quantum speed of the impact is, the higher is the intensity of the transmitted quantum impact. The electron process *blue shift* impact supply depends on the demand/consumption of these impacts on the other side of the balance. Space-Times exist in parallel.

The *quantum communication* of the internal and the "external" space-times does not eliminate the general objective of the pyramid to consolidate the two intensity levels inside and outside the pyramid, for connecting the *IQ* values of the centre and the surface.

Pyramid works like a quantum impact intensity (energy) pump.

There is a dominant (among the infinite number of) intermediate *IQ* drives from $IQ_{conflict} = \dfrac{c^2_{pxx}}{\varepsilon_{pyr}}$; to $IQ_{surface} = \dfrac{c^2_x}{\varepsilon_{pyr}} = \dfrac{c^2_E}{\varepsilon_E}$;

8E4

This is the *IQ* of the initial *blue shift* conflict with quantum speed $c_{pyr} = 383,870$ km/sec.

[*Gravitation* is about quantum impacts as well. The transformation between the *plasma* and the *Hydrogen* process creates elementary processes, which are not just for quantum speed reduction but also of modified electron process intensity – elementary processes with real anti-electron process *blue shift* quantum impacts.]

There is a need for having a receiver for utilising the quantum impact having been communicated by the *pyramid*. The "receiver" should be with space-time of the increased quantum speed equal to the one of the *pyramid* with external cooling.

Quantum impact is generating in the case of electron process *blue shift* conflict initiated by electron process surplus. This is valid only for the internal region with the highest quantum speed value. Other impacts are the result of the transformation towards the surface.

There are minerals with increased temperature if taken out at the surface of the *Earth*. The increased quantum speed of these minerals is the proof of their energy generation. The external *Earth* space-time is of less quantum speed than the internal space-time of these minerals.

There are here again two systems with highly differing "energy" intensity processes. But the difference is so significant that the increased impacts result in conflict, which generates heat. The best way is to cool it by water and follow the natural way, the function of oceans.

$$\frac{c^2_{mineral}}{\varepsilon_{mineral}} = n\frac{c^2_{water}}{\varepsilon_{water}} \quad \text{for reaching} \quad \frac{c^2_{Earth}}{\varepsilon_{Earth}} \quad \text{at the surface the water is steaming} \qquad \text{8E5}$$

In this case, the transformation happens in the water rather than the mineral!

The quantum speed of the mineral does not change. The quantum speed of the water is less than the quantum speed of the *Earth*. The water either is steaming on the surface or its volume (*n*) high enough for normal cooling. As an example, the most intensive heat producing elementary processes are the *Uranium* and *Thorium* processes, as expected with high quantum speed value close to the increased internal quantum speed of pyramids.

(*U238*: 9.46×10^{-5} W/kg and *Th232*: 2.64×10^{-5} W/kg).

[Source: Turcotte, D. L.; Schubert, G. (2002), Cambridge University]

Once the elementary structure of the minerals is losing on its space-time intensity, it will be communicating with similar space-times of equal intensity, taking intensity impact for re-establishing its intensity. This is an energy intensity transfer carried out by the quantum system above the *Earth* surface. The communicating quantum systems are of increased quantum speed, existing in space and time in parallel with the *space-time* of the *Earth*!

There is a simple way for receiving/utilising the energy of the internal *blue shift* conflict of pyramids.

S.
8.2

8.2
Intensity distribution

The intensity distribution of the quantum impact of *Earth gravitation* within the *pyramid* structure has a specific form. The quantum speed on the surface shall be equal to the quantum speed on the *Earth*. The increased *IQ* value within the pyramid is the result of the intensity difference between the surface and the internals. This is a consequence of the

Ref.
Fig.
8.2

difference between the surfaces of the casing and the base of the pyramid.
The intensity distribution of one of the segments of the structure is shown in Fig.8.3, proving the prediction in Fig.8.2.
The dimensions of the figure in Fig.8.3 are speed values.

Ref.
8A4

$$\text{With reference to 8A4:} \quad a = \frac{\pi}{2} r ; \quad \text{and} \quad r = \frac{2}{\pi} a$$

Ref.
8A3

With this size relation above, the surface related *Golden Ratio* is always valid. $\quad \varphi = \dfrac{S_{casing}}{S_{base}} = 1.6189$

In this way, the actual conflicting and dominant quantum speed of the internal quantum membrane of the pyramid can be of different value, function of the mineral composition, but the relation of the size parameters of the pyramid are one and the same. In other words, it can be stated that the length and the height values of the pyramid establish the acting quantum speed within, which generates the conflict and the quantum impact.

Ref.
8C12

With reference to 8C12, the conflicting and the normal quantum speed values represent the related *IQ* drives of the pyramid: $\quad \eta = \dfrac{c_{pyr}^2}{c_{gr}^2} = \varphi = 1.6189$

The division of the *pyramid* by segments below proves Fig.8.2 was correct.
The relations of the surfaces of the casing and the base are valid for each of the segments.

In this case, the intensity relation between the basis and the casing of each segment remains the same. Through the *apex*: $c_{pyr} = 383{,}870$ km/sec.	The length of the segments are of linear, the number of the impacts are quadratic function

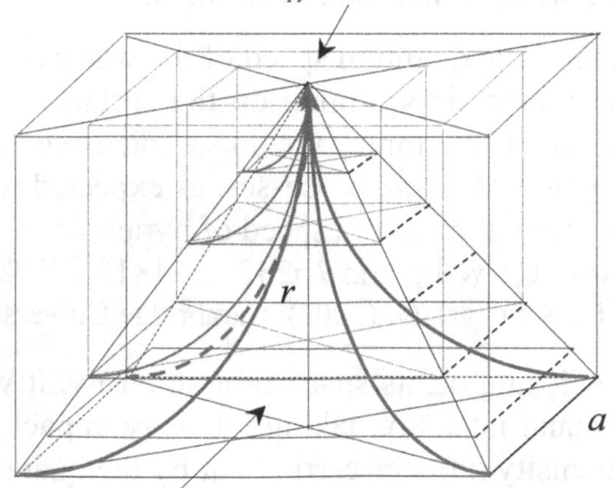

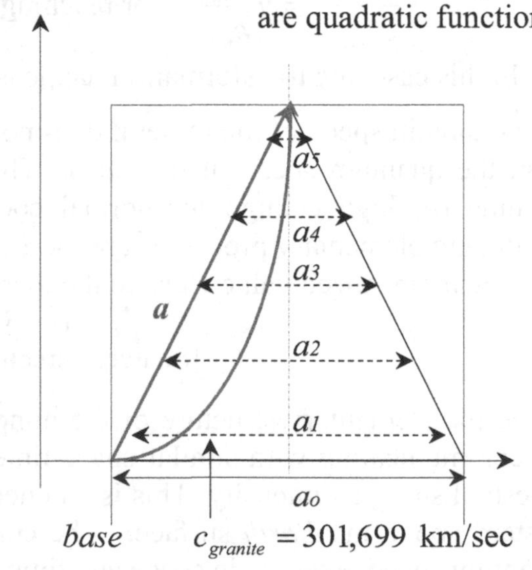

internal area of the *conflict* with *increased speed*

Fig. 8.3/a

base $\quad c_{granite} = 301{,}699$ km/sec

Fig.8.3/b

Fig.
8.3

The horizontal length of the base and the segments of the casing are proportional to the number of the acting quantum impacts of *gravitation*, speed value of $c_{granite}$.

With reference to 8A8 and 8C5, this number of impacts is the reason for the internal conflict. (This value of the quantum speed is for granite and it is valid only within the pyramid, taken for simplicity.)

Ref.
8A8
8C5

The speed of the quantum impacts for each of the segments has the same value, but the number of the impacts corresponds to the construction of the pyramid.

a_n the lengths of the base of the pyramid and of the segments are of linear function;

n_n the number of the acting quantum impacts of gravitation with speed $c_{granite}$ is a quadratic function as $n_n = a_n \cdot a_n$;

8F1

IQ_n the quantum drive is proportional to the number of the acting quantum impact of gravitation with speed $c_{granite}$,

$$IQ_n = n \frac{c_{gr}^2}{\varepsilon_{gr}}$$

8F2

as c_{gr}^2 is valid for all segments and ε_{gr} is constant.

There is an intensity distribution caused by the electron process blue shift conflict within the pyramid.

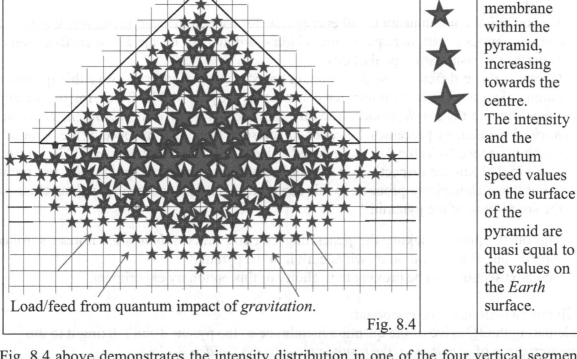

Load/feed from quantum impact of *gravitation*.

★ ★ ★ ★ ★ ★ Stars mark the intensity of the quantum membrane within the pyramid, increasing towards the centre. The intensity and the quantum speed values on the surface of the pyramid are quasi equal to the values on the *Earth* surface.

Fig. 8.4

Fig.
8.4

Fig. 8.4 above demonstrates the intensity distribution in one of the four vertical segments of the pyramid construction. Fig. 8.4 shows that the impact of gravitation also modifies the quasi homogenous intensity distribution of the *Earth* surface.

The layers of the *Earth* surface below the pyramid contribute to the building up of the quantum membrane within the pyramid in order to have fluent quantum communication with the quantum impact of gravitation.

There are here three important points to be noted:

1- *relativistic*: the time flow with reference to Section 5 within the pyramid is slower: events happen within the pyramid for shorter time, as the quantum speed within the pyramid is increased and equal to $c_p = 383,870$ km/sec.

Ref.
S.5

2- *thermodynamic*: the intensity increase as the result of *blue shift* conflict within the pyramid increases the temperature and generates heat.

3- *quantum impact related*: while the increased intensity of the conflict is decreasing towards the casing, there is a directed by the conflict high *IQ* value with $c_{pyr} = 383{,}780$ km/sec quantum speed towards the *apex* of the *pyramid!*

The quantum speed value of the constant quantum impact through the *apex* corresponds to the conflict, caused by the difference of the surfaces of the base and the casing.

In this way the energy source is given.

In order to utilise the energy intensity potential gift of the pyramid available, there is a need for a certain quantum impact intensity consumer.

As a principal conclusion there are here two options:

1. The generation of electron process *blue shift* impact intensity conflict and $c_p = 383{,}870$ km/sec quantum speed within the pyramid means:
   - the quantum communication of the pyramid with space-times of the quantum speed value, equal to the quantum speed of the pyramid; and
   - transfer of the generating quantum impact (energy) of the conflict to this space-time.

   This quantum communication and energy transfer happen through the global integrated quantum system, with our space-time within – without we on the *Earth* surface would register or measure any signal of this.

Ref.
Table
1.1
   Because of the difference in the quantum speed values, there is no way this quantum communication and energy transfer could be directly available for us for use in our space-time on the *Earth* surface. The quantum speed values of the *Uranium, Thorium, Radium* elementary processes, however, are close to the quantum speed of the pyramid. (With reference to Table 1.1 these speed values are 378,163 km/sec, 377,076 km/sec and 375,914 km/sec respectively.) The only way is for building up a station-reception using these elementary processes with quantum speed and space-time values close to the space-time of the pyramid.

2. The other option is taking the generating electron process *blue shift* impact surplus directly out of the pyramid by wires built into the pyramid.

   The next section will be proving the validity of this option by experiment.

A finding which might be important:

Calculating the *IQ* drive of the quantum membrane of the pyramid and relating it to the quantum drive of *gravitation*, the result is close to the *Golden Ratio*:

8F3
$$IQ_p = \frac{c_p^2}{\varepsilon_p} = \frac{383{,}870^2}{1.00421}; \qquad IQ_E = \frac{c_E^2}{\varepsilon_E} = \frac{299{,}792^2}{1}; \qquad \frac{IQ_p}{IQ_E} = 1.632$$

Note:

Before starting the next section with the results of the pyramid experiment, there are important annexes about the impact of *gravitation*, examined from the point of view of building structures, especially churches and the about the general characteristics of mountains – which might also be of interest. Presenting them here would be in conflict with the fluency of the main content.

Ref.
Annex
8.2
Annex
8.3

8.3
Pyramid is generating energy
Pyramid is a potential power plant
Results and the conclusion of the measurements with the *pyramid* replica

A replica of the *Great Pyramid of Giza* was built up for
- proving the internal electron process *blue shift* surplus and conflict of the pyramid construction, introduced in the previous sections on a theoretical basis;
- demonstrating the energy generating fundamentals of pyramids.

The replica is with base of 785.4 mm and height of 500 mm. The parameters of the *Great Pyramid* are 230 m of length and 146.5 m of (the original) height. The similar geometry relations mean the principle of the energy generation and the energy potential of the replica is similar to the *Great Pyramid*. The replica has been built up from a certain mineral composition, similar in their energy characteristics to granite, the building stone of *Giza*. Just that neither *Giza* nor the other *pyramids* have wires inside and a cable through the top, connecting their internal quantum membrane with the *Earth* surface.

As the quantum speed and the intensities of the replica and the *Giza* pyramid are the same, their generating voltage relation directly corresponds to their mass proportions. The proof is the definition of the electron process, the basis of the conflict: Only the mass values are different

$$\frac{dmc^2}{dt_i\varepsilon}\left(1-\sqrt{1-\frac{(c-v)^2}{c^2}}\right)$$ 8G1

The mass relation between the *Great Pyramid* and the replica is 25.78 million!

The measurements were made in the continental climate conditions of Hungary in March-April 2015.

Wires were built in the concrete structure with a measurement terminal connecting the internals with the soil for measuring the developing intensity voltage potential within the replica.

Two temperature indicators were built in at distance 1/3 of the height from the top and from the bottom
– for measuring the temperature increase, representing the internal conflict.
The temperatures at the surfaces at these heights were also measured.

The temperature of the air around the replica was measured.

Temperature measurements were
(1) on the surface of the soil,
(2) at 200 mm depth around the *pyramid* and (3) 200 mm below the *pyramid*.

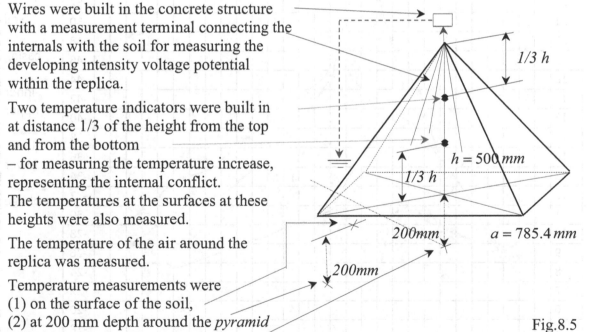

Fig.8.5

Fig.
8.5

The building material of the replica is a certain composition of minerals with elementary processes of *O, H, N, C, S, Ca, Si, Mg, Al, K, Na, Cl, Ti, Fe*. The unique characteristic of the mix of this mineral composition is its electron process *blue shift* surplus.
All measured data and pictures about the experiment are given in Annex 8.1.

Ref.
Annex
8.1

Day 25 March 2015 – without direct sunshine impact on the casing of the *pyramid*

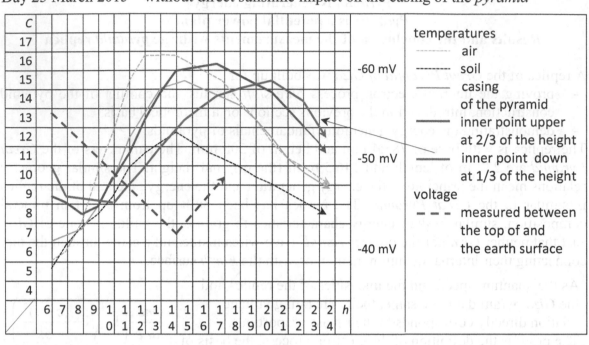

Diag.
8.1

Diag. 8.1

Day 26 March 2015 – with direct sunshine impact on the surface of the *pyramid*

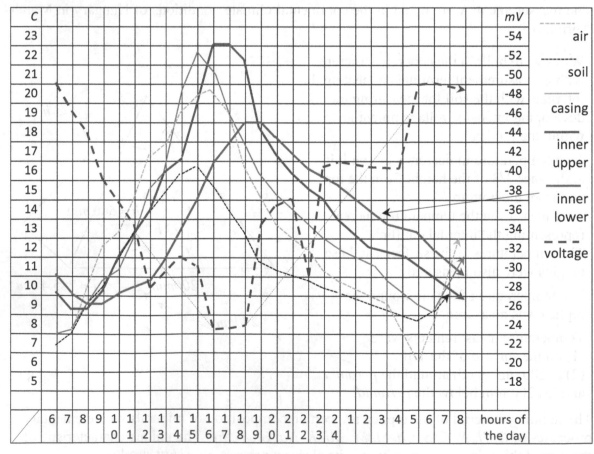

Diag.
8.2

Diag.8.2

All measured data are important, but the most important is the _minus voltage_ between the _Earth_ and the top of the pyramid replica.

Ref. Annex 8.1

Voltage measurement of any kind between these points is more than noteworthy, but having minus value here first of all raised concerns about the polarity of the connections. Without discussing all related controlling steps, the main arguments proving the measurement is correct are: the polarity is in order and plus voltage was also measured in different circumstances as will be presented later. This voltage is representing the potential difference, the intensity of the feeding quantum impact of _gravitation_ from below the pyramid towards the pyramid, as had been supposed within the theoretical section.

The temperature measurements within the soil were important, but not significant.
As the measured data in the Annex 8.1 prove, the temperature of the surface follows the temperature of the air, the temperature of the soil and the temperature below the pyramid differ, but their impact has no relevance.

The replica on both days was protected from the direct quantum impact of the _Sun_.
The external air temperature during the days was different comparing the measured data.
The pyramid is taking energy in from the quantum impact of _gravitation_ and the _Sun_. (Taking energy means: the anti-electron process _blue shift_ impact of _gravitation_, the direct quantum impact of the _Sunshine_ and the generating _blue shift_ conflict of the air heated by the _Sun_ are impacting the pyramid through the base and the casing surfaces respectively.)

Day 1 and 2 April 2015 – with cold wind and sunshine

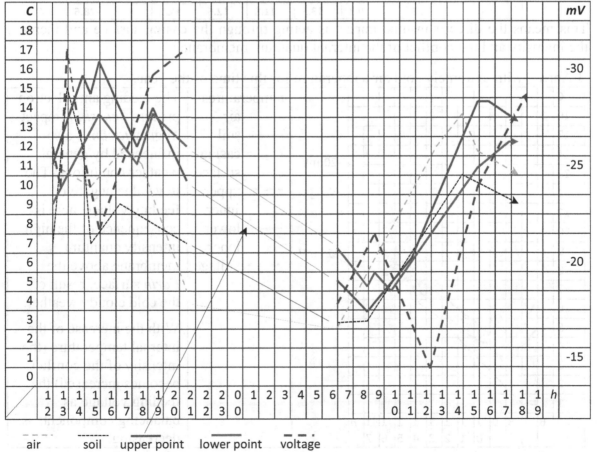

air   soil   upper point   lower point   voltage

Diag.8.3

The temperatures were different on these two days, but the results were similar:

-   for keeping the internal stability at a certain level, the change of the air temperature and the voltage of the feeding potential of gravitation are of opposite gradients.

The evening, late night and early morning data are missing from Diag.8.3. Temperatures can be estimated but the feeding impact of gravitation needs proof. Diag.8.5 of a different day gives the evidence: external temperature and gravitation impacts are of opposite gradients. For reaching the balance night periods are with increased gravitation impact.

With reference to the assessment of the measured data it can be concluded that it is not surprising that pyramids have been built up in hot places. For keeping the stability of the internal quantum membrane the feeding voltage from *gravitation* during night times should be of higher value in order to balance the intensity of the external air temperature impact during the day.

The data taken and the diagram of the presented two days' measurements with different external temperature clearly demonstrate:

the feeding from *gravitation* corresponds to the daily impact.

Diag. 8.3 also proves the internal temperature might be higher than the external temperature:

-   on 1. April both internal measured temperatures of the quantum membrane were higher all day, even during the hottest periods of the day – as this extract from Table 8.3 of the Annex demonstrates:

time	Air temp	Soil temp	Casing temp		Internal temp		Voltage
			up	down	up	down	
16:30	12.4	9.5	15.2	12.9	13.6	12.9	-25.7
17:30	11.3	9.3	13.5	11.8	12.6	11.8	-28.6

This means the direct sunshine impact is acting through the casing surface and increases the intensity and the conflict of the internal quantum membrane.

3 April – with extremely cold air temperature for this time of the year.

temperatures
------- air
-------- soil
——— casing surface
——— upper inner
——— lower inner
- - - voltage

The feeding anti-electron process *blue shift* impact of *gravitation* is also influenced by general weather conditions: The low value negative feeding voltage here is consequence of the high intensity quantum membrane above the *Earth* surface, result of the cold cloudy weather.

With the increase of the internal temperature (the intensity of the conflict becomes increased) the feeding voltage as balancing components is decreasing.

Day 26 and 27 April 2015 – with extremely high temperature

air   soil   upper point   lower point   surface   voltage

Diag. 8.5

Diag.8.5

With reference to Diag.8A1-8A5, the feeding gravitation quantum impact is decreasing with the increase of the internal temperature. Diag. 8.6 also proves the decreasing internal temperature initiates the increase of the value of the feeding voltage at night time.

Night of 6 to 7 April

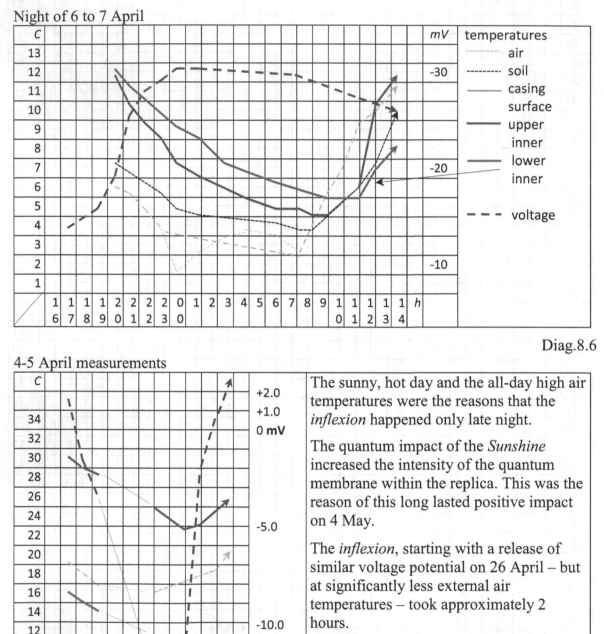

Diag.8.6

Diag. 8.6

4-5 April measurements

The sunny, hot day and the all-day high air temperatures were the reasons that the *inflexion* happened only late night.

The quantum impact of the *Sunshine* increased the intensity of the quantum membrane within the replica. This was the reason of this long lasted positive impact on 4 May.

The *inflexion*, starting with a release of similar voltage potential on 26 April – but at significantly less external air temperatures – took approximately 2 hours.
On 4 May it was about 12 hours.

The feed/load dominance of *gravitation* – as the diagram shows – was from 22:00 to 8:45.

temperatures: ----- air ——— inner upper ——— inner lower; – – – voltage

Diag. 8.7

Diag.8.7

The mains source of the feeding is *gravitation,* but in sunny, hot days the direct and the indirect *Sunshine* impact increases the internal electron process *blue shift* conflict of the quantum membrane of the pyramid. The *Sunshine* quantum impact partially "substitutes" the quantum impact of *gravitation*.

8.3.1 Conclusions

These are all important data for characterising the quantum energy balance of pyramids, but there is an extremely important finding registered during the measurements:

At the start of each voltage measurement between the top and the *Earth* the measured value was around *+200 mV*. This is the potential difference between the top – in fact, the internal quantum membrane of the pyramid – and the *Earth*. This potential difference during the measurement was always decreasing. The pyramid replica is releasing its internal potential into the *Earth*. The gradient of the change is about *–(0.1-1.0) mV/sec*.

The plus voltage means electricity release from the pyramid!

The releasing voltage impact, does not stop reaching *zero* value, rather it starts, building up an increasing negative feeding voltage with a gradient of low value. The feeding negative voltage through the same cable, connecting the *Earth* with the top, is stabilising at the value of the feeding impact of gravitation through the base surface.

The value of the stabilisation of the minus feeding voltage – as the measured data and the diagrams demonstrate – depends on the given weather conditions of the day.

Any movement of the negative end of the voltage measurement during the potential release of the pyramid – any movement of the end, which is contacting the *Earth* surface – results *in constant plus voltage generation*, value of around *250-200 mV*. Moving this contact in any direction generates the renewal of the release starting again and again from the highest potential value. (The movement of the other end does not have any impact.)

With the change of the contact to the surface the released voltage was reaching *820 mV*.

The pyramid generates energy! Better to say: the pyramid is transferring the quantum impact of gravitation into measurable voltage potential!

If the mass proportions would be linear, with the experienced generation of around *200 mV* by the replica of 500 mm height, the supposed generation of the standard *6 kV* potential would need height of *7.82 m*. But with reference to Section 8.4 the measured voltage represents only the potential difference between the two ends of the wire.

Important findings:
1. There are two directions of the electron process *blue shift* impact flow (direct current) within the wire connecting the top of the pyramid and the soil of the *Earth* surface: with *positive* voltage is the DC towards the *Earth* and with *negative* voltage is the DC from the *Earth* surface towards the top of the *pyramid*.
    -   the first means the *pyramid* is giving off *blue shift* conflict intensity (electricity),
    -   the second means *gravitation* feeds the *pyramid* (by the anti-electron process *blue shift* impact of gravitation) through the top.
2. The *pyramid* is with electron process *blues shift* surplus and conflict. Connecting the top of the *pyramid* with the *Earth* surface with a cable, the intensity surplus is releasing from the quantum membrane in conflict within the pyramid. The replica pyramid is producing a fluent intensity (+DC) release with a potential starting from around of *200-250 mV* or even from *820 mV* (with *Al* alloy contact to the *Earth*) to *0* mV.
3. The release of the +DC voltage electricity potential does not stop at the inflexion point with $U = 0$. The energy (+DC) release turns into energy feed – anti-electron process *blue shift* impact flow (– DC).

In their physical quantum impact there is no difference between electron and anti-electron processes. Both are causing *blue shift* impact. The negative DC flow from the *Earth* surface, however, needs more explanation:

The internals of the pyramid have been in electron process *blue shift* conflict of increased intensity. The intensity of the anti-electron process *blue shift* impact (the energy intensity feed/load of *gravitation*) from the *Earth* surface around the *pyramid* is less than the intensity below the *pyramid*. The reason is the difference between the surfaces of the casing and the base of the pyramid. The intensity of the quantum impact of gravitation around the pyramid is equal to the intensity of the quantum impact of gravitation through the casing of the pyramid.

The pyramid has been under permanent load through the base in line with the continuity of the quantum impact of *gravitation*. The intensity of the quantum impact of *gravitation* through the base is higher than the intensity through the casing. This results in electron process *blue shift* conflict within the pyramid, creating a quantum membrane of increased intensity. The conflict within the pyramid also means increased quantum speed. The wires built within the pyramid have been impacted by the electron process *blue shift* conflict of the quantum membrane. The elementary process of the wires does not need extra electron process *blue shift* impact. Therefore, the surplus has been released through the cable from the top to *Earth*.

The wires and the cable of the release, connecting the top and the soil are the tools of the quantum communication between the *Earth* and the internals of the pyramid. The intensities of the two endpoints are corresponding at one side to the internal conflict of the pyramid, and, on the other, to the intensity of the quantum impact of gravitation (of normal not increased value).

The cable is connecting these two intensities. This is the reason the quantum impact of *gravitation* from the *Earth* surface (of normal intensity) will be impacting the internals through the wire after the inflexion as well. This is the reason the release turns into feed after the inflexion.

The measured minus anti-electron process feeding voltage is representing the intensity of the gravitation impact through the basic surface. (The feeding through the base is increased in numbers rather than in intensity.)

Without the disruption of the connection between the top and the *Earth* surface, the feeding anti-electron process *blue shift* quantum flow, the minus current (–DC) is on and is contributing to the integrated quantum feed of the *pyramid*.

4. If the feeding contact is disrupted, the feed stops and the release takes over. And the process continues as written above.

5. If the contacting end of the release cable is being permanently moved on the *Earth* surface, the release starts again and again. Each position on the *Earth* surface means a new quantum condition. As the intensity of the pyramid internals are higher than the intensity of the feeding impact of *gravitation* the release is obvious and starts from the highest intensity again and again. As the experiment clearly proves.

6. The intensity loss of the release is compensated by the quantum impact of gravitation through the base.

Whatever the value of the release is, the load from below has been guaranteed by gravitation.

The potential difference, the feeding impact depends on the following factors:

a. the intensity of the quantum system above the *Earth* surface

Meaning: cloudy weather conditions, for example, mean increased quantum membrane above the *Earth* surface – the feeding is limited and is about the potential of around -20 mV (as Diag.8.3 and 8.4 demonstrate).

b. the external temperature

Meaning: the internal energy potential is not just the result of the quantum impact of *gravitation* and the quantum impact of the *Sunshine* but (as consequence) has also been influenced by the external temperature. Higher temperature is causing the increase of the internal conflict. The minus feed is decreasing during the day. The low temperature at night time and early mornings initiates high minus feed. Diag.8.1, 8.2, 8.3 and 8.5 demonstrate both.

With the decrease of the internal *blue shift* conflict of the quantum membrane the feeding supply towards the pyramid is increasing. The lowest daily voltage supply is at the highest intensity of the internal conflict, corresponding to the highest internal temperature.

Additional important factors:

1. The internal conflict is generated by the quantum impact of *gravitation*.

The intensity of the quantum impact of gravitation through the casing is less than the intensity through the base of the pyramid. This results in internal electron process *blue shift* conflict.

2. While the *Sunshine* as direct and indirect "external" quantum impact increases the conflict and by that the temperature, the main motivating source is and still remains *gravitation*.

Diag. 8.1 and the respective data in Table 8.1 taken on 25 March 2015 are with no direct *Sunshine* impact. (The replica was protected from the *Sun*.) The temperatures of the inner points are higher than the temperatures of the casing. It clearly means the main source of the heating impact is the quantum impact of gravitation.

Diag. 8.2 was especially arranged for measuring the case having direct *Sunshine* impact. In this case, the separation of the two impacts (*Sunshine* and *gravitation*) is more difficult, but the temperature of the inner upper position at 16:00 hours and 17:00 hours is clearly higher than the temperature at the surface. The internal temperature at the upper point of the pyramid replica of the 26th April measurement (Diag.8.5) was all day higher than the temperature of the surface. (In heating-cooling relations, the temperature of the heated side can never exceed the temperature of the heating source, not even in the case of the accumulation of the heat.) So, the basic energy source is not the sunshine coming through the casing. The diagrams prove that this is the anti-electron process *blue shift* impact of the elementary processes of the elementary evolution – *gravitation*.

3. The temperatures of the soil around the *pyramid* and below the *pyramid* were always less than the measured temperature within the *pyramid*.

It directly means that the temperature of the soil has no increasing impact on the internals (the internal conflict) of the *pyramid*. The close to the pyramid base regions of the soil could not be measured and below at 200 mm level there was no impact.

4. The measured feeding voltage in normal weather conditions is decreasing during the day and increasing again approaching the evening and night as the external temperature is decreasing.

Ref.
S.
5.2.3

5. Keeping the cable connection between the top of the replica and the *Earth* surface with the voltage measuring device turned off, the feeding is going on. The proof is that if the measuring device is on again, the measured value corresponds to a negative feeding impact. If the cable connection with the surface, however, is interrupted the measurement starts with the measurement of the release of the intensity potential (positive voltage). *This proves: the quantum communication between the Earth and the internals of the pyramid is going on even with a turned off measuring device.*

6. The electron process *blue shift* conflict within the *pyramid* is resulting in quantum speed increase, establishing a space-time of increased intensity.

7. The release of the voltage from the replica is always positive.
   It represents the electron process intensity potential of the pyramid independently on the actual dominant source of the load (*gravitation* or *Sunshine*).

   If the release of the positive voltage, going through the step by step decrease and approaching the *inflexion* changes to negative value = it means the release switches to load through the same cable connection. The negative voltage demonstrates the load from *gravitation*.

   If the value of the release is decreasing step by step but its value does not change from positive to negative = it means the release can be compensated by the electron process *blue shift* conflict, accumulated within the quantum membrane of the pyramid. This does not mean at all that the load from *gravitation* is missing, just the fact that the resulting potential between the pyramid and that certain point of the *Earth* surface is still positive.

8. With the permanent move of the point contacting the *Earth* surface with the top of the pyramid, the intensity release is restarting again and again from its maximal value. The energy release this way becomes constant. The internal balance of the pyramid cannot be disrupted and the intensity loss is compensated by the acting feeding sources!

9. The intensity of the quantum impact of *gravitation* on the *Earth* surface and below the pyramid is the same, just through the base of the pyramid in higher numbers. Therefore the cable contact between the top of the pyramid and the *Earth* is well representing the intensity relations below and inside the pyramid.

   If the intensity of the quantum membrane of the pyramid is higher, the measurement is with positive voltage. If the intensity within the pyramid is less, the measurement is the opposite: with negative voltage, representing the electron process supply from the *Earth* to the pyramid.

8G2 $$\Delta_{measured} = -e_{gr} + e_{Sun} \quad \text{if } |e_{Sun}| > |e_{gr}| \text{ the measured value is positive;}$$

$$|e_{Sun}| < |e_{gr}| \text{ the measured value is negative.}$$

e_{gr} - is the anti-electron process flow towards the pyramid, representing the feeding impact of *gravitation* through the basic surface;

e_{Sun} - is the electron process flow towards the *Earth*, representing the intensity potential of the pyramid, source of air and *Sunshine*.

The flow in both cases is electron process, just the directions are different.

The measured data and pictures about the experiment are presented in Annex 8.1 in details.

8.4
The quantum impact of *Earth gravitation* is transported to a different *space-time*

Without the wires built in, the generating electron process potential (energy) is transported by the generating quantum speed of the pyramid through the all over quantum system to a specific "place", a *space-time*, with the quantum speed of this *space-time* equal to the quantum speed of the pyramid.
The quantum impact of Earth gravitation is transported to a different space-time!

The wires built in, well illustrate the generating voltage. But the wires connect the internals of the pyramid with the *Earth* surface, with its $c = 299,792$ km/sec quantum speed *IQ* potential. This *Earth* potential in this case is the one establishing the demanding end.

The electron processes of the wires are communicating with both ends, as all electron processes have the same dt_i time system. But the developing voltage in fact is the potential difference between the two ends of the wire (only). The experience demonstrates, that different materials, as contacts on the *Earth* surface develop different voltage.

The quantum impact of the electron processes of the conflict within the pyramid generates electron process *blue shift* impact within the wire.
> The conflict, generated by the quantum impact of the anti-electron process of gravitation is a kind of surplus. The surplus is impacting the elementary process of the wire. But the wire is in balanced status. This way the impact goes through the wire and it ends as being "swallowed" by the *Earth*, as the *Earth* is surplus potential is less than the conflict within the pyramid.

The space-time of the pyramid is connected *to another space-time* (the really acting end) with quantum speed value, equal to the quantum speed of the internal conflict of the pyramid.
Space-time quantum communication is distance-independent, function only of the speed value of the quantum communication.
There are two conditions, which cannot be broken:
  - the continuity of the quantum impact of *gravitation* through the casing of the pyramid;
  - the continuity of the quantum communication between the space-times of equal quantum speed.
The other space-time is developing the demand by the utilisation of the transported energy. The pyramid is covering the demand by transferring the quantum impact of *gravitation* through the overall quantum space. No use means no any balance deviation – no transport. Pyramid works as a "remotely controlled energy pump".

8.5
Sunshine, the quantum impact

The quantum system of the *Sun* means a *space-time* of higher quantum speed and intensity. The genuine quantum speed of the beam of the *Sunshine* is higher than our quantum speed on the surface of the *Earth*. The conflict with the quantum impact of the *Earth* slows down the speed and causes conflict. The conflict results in heat generation and in the intake of the slowed down quantum impulse (energy).

The quantum speed of the *Sunshine* beam is very likely close to the quantum speed of the pyramid of the classical configuration of the *Giza*, generated by the internal conflict. This is coming from the experience of their direct communication.

The *Sunshine* is impacting the internal quantum membrane of the pyramid without and bypassing the increase of the external air temperature.
We do not intercept the direct quantum impact of the *Sun* on the surface of the *Earth*. We experience only the consequence of the conflict.

The overall quantum space is full with infinite number of quantum signals.
The interference of signals always results in conflict. Conflicting quantum impacts and signals lose on their intensity. In the case of the *Sun*, the density and the intensity of the impact are the ones resulting the overall conflict and heat generation. The density of the impact in different periods of the year is different. The direct quantum impact to the pyramids is without change. This is the conflict and the heat generation, which are impacted by the density.

ANNEXES

Old Testament – Genesis

"The Creation
(1) In the beginning God created the heaven and the earth.
(2) And the earth was without form and void and darkness was upon the face of the deep. And the spirit of God moved upon the face of the waters."
These are the first lines of the First Book of Moses of The Old Testament: *Genesis*.

Elementary evolution starts at the *plasma* state and ends with the *Hydrogen* process.
Plasma is neutron collapse with *inflexion* to the anti-neutron expansion, both of infinite high intensities, generating anti-electron process *blue shift* surplus of infinite large volume. The *Hydrogen* process – in the contrary – is the whole expansion with the electron process *blue shift* drive and the collapse of the neutron process of infinite low intensities.

The evolution process from the *plasma* to the *Hydrogen* process produces infinite number of elementary cycles. The elementary processes generate infinite number of quantum impulses (*quantum*). The generating *quantum impulses* establish the quantum systems of space-times.

The space-time of the *plasma* is with infinite high quantum speed and is of infinite high intensity.	The space-time of the *Hydrogen* process is with infinite low quantum speed and of infinite low intensity.

The elementary evolution from the *plasma* to the *Hydrogen* process integrates all elementary processes. The complex integrated status is the *Earth*, with the *plasma* in the centre with infinite high intensity of the expansion and with the *Hydrogen* process on the periphery with the fully expanded state of infinite low intensity.

The anti-electron process *blue shift* surplus of the elementary processes of the evolution generates the quantum impact of the expansion – *gravitation*. The expansion from the *plasma* state of infinite high intensity to the *Hydrogen* process of infinite low intensity results in the sphere symmetrical expanding acceleration of the *Earth*.

The quantum speed of the *Hydrogen* process is of infinite low value.
There are also other elementary processes close to the end stage of the evolution with natural quantum speed and intensity less than the formulating quantum speed and intensity on the external surface of the *Earth*. One of these is the *Oxygen* process.
As result of the expansion the quantum speed and the intensity is changing from the infinite high value of the *plasma* to the infinite low quantum speed and intensity value of the *Hydrogen* process.
For balancing the fluent transition from the infinite high quantum speed and space-time of the *plasma* to the infinite low quantum speed and intensity of the *Hydrogen* process and for the continuity of the expansion, the elementary communication close to the *Earth* surface region produces *water*, an elementary mix of processes, result of the quantum communication of the *Hydrogen* and the *Oxygen* processes.

The *water* covering the *Earth* surface connects
- the *Hydrogen* process out of the *Earth* structure, with quantum speed and the space-time of infinite low value; with
- the elementary processes within the *Earth*, with quantum speed and space-time of increased intensity – further increasing towards the centre with *plasma*.

and ensures the fluent transition of the quantum speed and the space-time between the *Hydrogen* process and the other elementary processes integrated within the *Earth*.

Water is the "instrument" connecting those above and below the *Earth* surface.
Exactly as the *Genesis* reads.

And the first lines further in the *Genesis* continue:
"The First Day:
(3) And God said: Let there be light: and there was light.
(4) And God saw the light and it was good: and God divided the light from darkness.
(5) And God called the light Day and the darkness he called Night. And the evening and the morning were the first day."

The quantum impulses generated by the elementary cycles establish the quantum system above the *Earth* surface. The sphere symmetrical expanding acceleration and the constant quantum impact of *gravitation* loads it. There is a *quantum membrane* above the *Earth* surface covered by *water*.

This *quantum membrane* receives electron process *blue shift* impacts from the *Sun* and the light is "turned on" this way indeed. The constant *blue shift* impact and the *blue shift* conflict from the *Sun* however are not just move and rotate the *Earth* but also form the *Earth* into spherical shape.

(*Sun* is in similar *plasma* state, with generating electron process *blue shift* impacts, but of increased quantum speed. The increased quantum speed and the increased intensity generate conflict within the quantum membrane above the *Earth* surface.)

So, there was light and darkness and evening and morning indeed.

Here the point is on the sequence: first obviously was the evening, as the first day ended by the rotation of the *Earth*, result of the *blue shift* conflict of the quantum impact of the *Sun* and the *quantum membrane* of the *Earth*.

And *Genesis* goes on:
"The Second Day:
(6) And God said, Let there be a firmament in the midst of the water, and let it divided the waters from the waters.
(7) And God made the firmament, and divided the waters which were under the firmament from the waters which were above the firmament: and it was so.
(8) And God called the firmament *Heaven*. And the evening and the morning were the second day."

With the progress of the elementary evolution, the elementary process structure of the mantle and the close to the surface regions of the *Earth* become more and more complex with more and more elementary processes, with more and more anti-electron process quantum impacts. The quantum impacts do not just increase the intensity of the *quantum membrane* of the water, but the increasing momentum of the elementary evolution is bringing up more and more elementary processes to the *Earth* surface.

"(9) And God said, Let the waters under the heaven be gathered together unto one place, and let the dry land appear: and it was so."

The increased in numbers quantum impacts (*gravitation*) generate electron process *blue shift* conflict within the *water*. Resolving the conflict water partially is steaming up leaving the liquid status.

The Old Testament in fact gives the formation of our planet.
The only way for being in harmony with the statements of this many thousands of years old wisdom is looking at the world on process basis.

**Annex
2.1**

Spiritual meaning of the time count difference
(Attachment to Section 2.3.3 on the intensity difference of elementary processes)

Elementary balance, process and anti-process relations, intensities, quantum speed, space-time and quantum communication can be applied to our human relations as well.

There have been two personal space-times taken with speeds values of quantum communication c_A and c_B. These speed values represent the integrated quantum speed of the elementary processes of *Person A* and *Person B*.
Both persons have been working, dealing with, are part of the same event v.
The question is what is the relation of their personal space-times?

2J1

$$\Delta t_x = \frac{\Delta t_o}{\sqrt{1 - \frac{v^2}{c_x^2}}}$$

The time formula gives the personal time flow (the time spent for the event) by both *Person A* and *Person B*.

2J2

It is taken that $c_A > c_B$; the used personal time for the same event for *Person A* is less: $\Delta t_A < \Delta t_B$

It is also taken, that $\Delta t_A < \Delta t_E < \Delta t_B$ - the personal time, spent by *Person A* is less, the personal time for *Person B* is more than the general time count on the *Earth* surface.
The results in 2J2 also mean that the speed of the quantum communication of *Person A* is higher and of *Person B* is less than the quantum speed of the *Earth*.
Δt_o is the time system of the *inflexion* of relative rest within our space-time – equal for all three cases.

There are here three space-times existing in parallel:
The space-time of *Person A* with speed value of the quantum communication c_A; *Person B* with c_B; and the space-time of the *Earth* with c_E. One is more and the other is of less

2J3

intensity than the intensity of the space-time on the *Earth* surface: $c_A > c_E > c_B$

Whatever is the counted time (duration) of the event within the space-time of the *Earth* the personal space-time depends on the personal speed values of quantum communication.
The measured duration of the event within the space-time of the *Earth* for each of the persons has secondary importance.

**Ref.
2J1**

The duration of the event in the *Earth* space-time is Δt_E and while formally this is also the duration of the event for both participants, the time counts, with reference to 2J1 above are different, shorter and longer for *Person A* and *Person B*.
The intensity of the event is higher within the space-time of *Person A*.
The "aging", caused by the event for *Person A* is less. *Person A* uses less personal life-potential for the event than for *Person B*.
The reason is the higher quantum speed of communication.

The absolute work values for both persons are equal and the same. For managing this work *Person A* however has used less personal time than *Person B*. It means the efficiency of the elementary operation of *Person A* is more; his aging is with lower gradient.

Additional thoughts on processes and anti-processes

The intensity of the electron process is established by the intensity of the proton process.

The intensity of the anti-proton process, the balancing/controlling part of the proton process is the one determining the intensity of the proton process: changing this way the elementary process to another one with reduced proton process intensity.

(As part of the natural elementary evolution the anti-proton process becomes of reduced intensity – result of the expired for *gravitation* anti-electron process *blue shift* impact and of the entropy of the cycle.)

Plasma is *blue shift* conflict and the end-result of the electron process *blue shift* conflict of infinite high intensity.

Plasma is releasing *blue shift* impact in the form of anti-electron process *blue shift* of infinite high intensity. The reason of the *blue shift* conflict itself is the electron process *blue shift* drive of infinite high intensity. The anti-direction is the way releasing this intensity surplus.

The *inflexion* itself is a change for $dt = 0$ time. The *inflexion* as elementary change for *zero* time happens in all elementary processes of any time systems. Therefore the *inflexion* as status of rest has its relativistic meaning: dt_o the time system of "rest" means the time system of the elementary process where the *inflexion* itself happens!

The "shortest" in this meaning relativistic time system (with the intensity of infinite high value) where *inflexion* happens is the *plasma* state: Collapse of infinite high intensity with anti-electron process generation of infinite high intensity.

There is however an absolute rule for all *inflexions* without exemption: independently, whatever is the relativistic time system the duration of the change is *zero*!

By cooling, *plasma* is approaching its full expanded status at the *Hydrogen* process state with infinite high proton process potential and electron process *blue shift* impact of infinite low intensity, but in conflict again: The neutron process of the *Hydrogen* process is of infinite low intensity and the accumulating proton process potential is trying to transform all developing intensity reserves into electron process *blue shift* impact.

Hydrogen and *plasma* are the two ends, the two sides of the elementary quantum communication connected by *space* and *time*.

<u>*Important note*</u>:
Changes are not self-destructive.

Anti-processes ensure and control the continuity of the elementary progress.

There is no way for changing the quantum speed from $v = \lim c = \infty$ to $v = \lim c = 0$ as of an immediate step.

Anti-processes provide all necessary elementary impacts for "controlling" the elementary progress in normal way. Processes and anti-processes happen in parallel. There is no need for their separate definition. They belong to each other and shall be understood under the definition of *process*.

Anti-processes play significant stimulating effect in elementary quantum communication. With the use of the *plasma* potential, as the cooling by *gravitation* goes on, anti-processes in the background initiate elementary evolution, with all necessary elementary corrections. The equation of the elementary balance is:

3G1
$$\frac{dmc^2}{dt_p \varepsilon_p}\left(1 - \sqrt{1 - \frac{i^2}{c^2}}\right) = \xi \frac{dmc^2}{dt_n \varepsilon_n}\sqrt{1 - \frac{(c-i)^2}{c^2}}\left(1 - \sqrt{1 - \frac{i^2}{c^2}}\right);$$

3G1 can also be written as:

3G2
$$\varepsilon = \frac{\Delta t_n}{\Delta t_p} = \xi \frac{\varepsilon_p}{\varepsilon_n}\sqrt{1 - \frac{(c-i)^2}{c^2}}; \qquad \text{or} \qquad \gamma = \frac{\varepsilon_n}{\varepsilon_p} = \xi \frac{\Delta t_p}{\Delta t_n}\sqrt{1 - \frac{(c-i)^2}{c^2}}$$

ε – is for the characterisation of the *elementary processes*; γ – is for *gravitation*.

3G4   The two together give the completeness of the elementary processes: $\varepsilon \cdot \gamma = \dfrac{\Delta t_n}{\Delta t_p}\dfrac{\varepsilon_n}{\varepsilon_p} = 1$

3G5   The balance also gives the answer to the value of the *quantum entropy*: $1 = \xi\sqrt{1 - \dfrac{(c-i)^2}{c^2}}$

3G6
As
$i = \lim a\Delta t = c$; and
$\lim(c - i) = 0$
$$\lim \xi = \lim_{\lim i = c} \frac{1}{\sqrt{1 - \frac{(c-i)^2}{c^2}}} = 0$$
the value of *quantum entropy* is approaching zero from above indeed !

As summary of the elementary background function of the process/anti-process relations
- the infinite high intensity of the *blue shift* drive of the *plasma* is resulting in collapse of infinite high intensity;
- the infinite high intensity of the collapse cannot be without controlling anti-event;
- the infinite high intensity of the *plasma* collapse also generates expansion as anti-process;
- the infinite high intensity of the collapse of the *plasma* is result of the infinite high *blue shift* conflict of all expiring proton process intensities from all elementary cycles;
- but cycles cannot be with one end only; there is no way having only neutron collapse without proton expansion;
- at one end the neutron process is of infinite high intensity, on the other end the proton process is of infinite low intensity;
- this can be formulated also in different way: at one end the drive of the collapse is of infinite high intensity, while at the other the anti-drive is of infinite low intensity;

How can the proton process develop infinite high driving intensity at one end and infinite low intensity at the other?
- the event at one end is of infinite high intensity (resulting in collapse of infinite high intensity) while the original quantum speed and the intensity of the event, both are of infinite low values; but as direct consequence, the *space-time* of the event is of infinite small value, resulting in infinite high conflict and collapse of infinite high intensity;
- on the other end the original quantum speed and the intensity of the event (*plasma*) is of infinite high values; but as consequence with infinitely expanded space-time and with infinite high volume of the anti-electron process *blue shift* surplus the anti-electron process is with infinite low intensity – controlling the case and driving the anti-proton process by infinite low intensity
- at *plasma* state the anti-neutron process is of infinite high intensity, the anti-proton is of infinite low intensity.

This infinite high intensity is consequence of the infinite high intensity of the electron processes, the direct and the anti-electron processes as well.

Blue shift impacts are *blue shift* impacts from either ends: intensities of the direct and equally of the anti-processes as well.

- the *plasma* as *inflexion* of infinite high intensity would be with permanent infinite high intensity – a change in fact without change, which needs control process;
- the intensity of the anti-electron process drives the anti-protons into collapse, the intensity of which in parallel shall correspond to the intensity of the proton processes covering the neutron process and providing the infinite high electron process intensity of *plasma*;
- the not used for anti-proton collapse anti-electron process *blue shift* impact is taken by *gravitation*.
- the proton process provides the electron process drive and cover to the neutron collapse;
- the *blue shift* impact of the anti-electron process drive, source of *gravitation* is not used for anti-proton collapse, since the proton process close to the *plasma* is of infinite low value, step by step increasing;
- the space-times of the elementary evolution are acting in parallel:
   the space-time of the *plasma* is of infinite high intensity;
   the space-time of the *Hydrogen* process is of infinite low intensity.

There are here two consequences of *gravitation*, the cooling of *plasma*:
- decreasing tendency of the electron process *blue shift* impact, the drive of the neutron process – elementary evolution;
- the quasi not used anti-electron process *blue shift* impact is the source feeding *gravitation*;
- the *gravitation* part of the anti-electron process impact starting from the *plasma* state and ending with the *Hydrogen* process is less and less; (in the case of the 8 elementary processes with proton process dominance the *gravitation* impact is missing)
- the intensity of the proton process is a relativistic value, since the intensity of the proton process is quasi constant, but in parallel the intensity of the neutron process is decreasing as the intensity of the electron process is decreasing.

The other end is establishing:
- neutron process with infinite low intensity, with in fact no anti-electron process *blue shift* impact as the electron process *blue shift* impact, collapsing the neutron process is also of infinite low value;
- there is a point where the potential of the *gravitation* part of the *plasma* anti-electron process *blue shift* impact comes to its end.

Natural isotopes

The intensities of the processes at both ends of the elementary cycle (at the *inflexions* of the neutron/anti-neutron and the anti-proton/proton processes) are equal:
- the intensity of the collapse of the neutron process determines the intensity of the expansion of the anti-neutron process;
- the intensity of the collapse of the anti-proton process determines the intensity of the proton process expansion.

Collapses have been driven and the *inflexion* transmits the intensity.

3H1   If the intensity of the neutron collapse is of $e = \dfrac{dmc}{dt\varepsilon}\left(1 - \sqrt{1 - \dfrac{(c-i)^2}{c^2}}\right)$;

where ε the intensity coefficient of the electron process clearly reflects the intensity relation of the proton and the neutron processes (characteristic of the elementary process),

3H2   the intensity of the anti-electron process will be: $e = \dfrac{dmc}{dt\varepsilon_{-m}}\left(1 - \sqrt{1 - \dfrac{(c-i)^2}{c^2}}\right)$;

3H3   where $\varepsilon = \dfrac{dt_n}{dt_p}$; and $\varepsilon_{-m} = \dfrac{dt_{-p}}{dt_{-n}}$;   and $\varepsilon = \dfrac{1}{\varepsilon_{-m}}$

ε_{-m} means the intensity of the anti-electron process, modified by the load to *gravitation*.

3H4 $$\varepsilon_{-m} = \varepsilon_- - \varepsilon_g$$

In the case of neutron process dominant elementary processes ($\varepsilon < 1$) the anti-proton process is driven by less intensity and results in an anti-proton/proton *inflexion* of less intensity. The proton process obviously becomes of intensity, less than that of the neutron process therefore generates electron process drive of increased intensity.

If $\varepsilon > 1$, meaning proton process dominance and electron process surplus: the anti-neutron process generates anti-electron process of increased intensity – in order to drive the anti-proton process of increased intensity.
Elementary processes of this kind also have distinguishing characteristics:
- apart of the *Hydrogen* process all of them have intensity coefficient right above value 1.
- and communicate with elementary processes of $\varepsilon < 1$ intensity coefficient.

It is important to note that the elementary evolution from *plasma* to the *Hydrogen* process is a fluent process. Proton and neutron processes of elementary processes happen within reasonable limits of "from"-"to". Therefore the intensity coefficients for an elementary process may cover wider range than just belonging to a certain proton "weight". As the diagram below demonstrates it – for the elementary processes of the evolution close to *plasma* and close to the *Hydrogen* process only.

There are plenty of natural isotopes within the chain of the elementary evolution, where the intensities of the proton and the neutron processes do not correspond to the "standards of the elementary process".
Natural isotopes, developing within the chain of the natural elementary evolution do not cause any complication or conflict or destroying effect. The imbalance is vital part of the natural change. Natural isotopes mean the intermediate steps between stable balanced elementary structures.
Therefore natural isotopes can be found when mining for minerals.

The consolidation/rehabilitation of natural isotopes, taken out of the *Earth* mantle is difficult: The reason is that the overall elementary anti-process impact, acting within the *Earth* from the *plasma* on, as source of the rehabilitation disappears. Natural isotopes on the surface of the *Earth* are also with harming effect, as the protective *Earth's* layers are missing.
Isotopes are also generated in the case of the destruction of the balance of the elementary processes.

...

Plutonium process: (94)

$$^{244}Pu \rightarrow {}^{239}Pu \rightarrow {}^{238}Pu$$

Neptunium process: (93)

$$^{239}Np \rightarrow {}^{237}Np \rightarrow {}^{235}Np$$

Uranium process: (92)

$$^{238}U \qquad \rightarrow \qquad {}^{232}U$$

Protactinium process: (91)

$$^{234}Pa \quad \rightarrow \quad {}^{231}Pa \rightarrow {}^{229}Pa$$

Thorium process: 90(

$$^{234}Th \rightarrow {}^{232}Th \qquad \rightarrow \qquad {}^{227}Th$$

...

Helium process: (2)

$$^{4}He \rightarrow {}^{3}He$$

Hydrogen process: (1)

$$^{3}H \rightarrow {}^{1}H$$

Beta (minus) and *gamma* radiation are with especially damaging impact.

Both are results of the not sufficient electron process *blue shift* drive of the neutron process.

The rehabilitation of these isotopes is possible by the quantum impact of other elementary processes with proton process dominance. The balance improving effect of the damaged elementary processes is as follows:

1. Proton process dominant elementary processes have electron process *blue shift* surplus;
2. and the available surplus can be used for driving the neutron processes of the damaged elementary process of the isotope.
3. Neutron processes are neutral, (with reference to Section 1, $c_x \cdot \varepsilon_x = const$ for all elementary processes and the time system of the communication is quasi $dt_i \cong const$);
4. therefore the neutron process of all elementary processes can be driven by the electron process *blue shift* drive of other elementary processes.
5. The cycle of the elementary process in this case corresponds to the one of the drive, which is with balanced elementary process.
6. This is equivalent to the decrease of the intensity demand of the direct electron process of the isotope.

Alpha isotopes are with proton process of increased intensity, *beta (plus)* isotopes are with additional electron process intensity surplus. The rehabilitation of these isotopes is easy as the natural defects expires.

The rehabilitation of isotopes is examined in details in Section 7.

**Annex
4.1**

Quantum communication in practice

Quantum communication of elementary processes is based on
- the quantum speed,
- the intensity of the electron processes,
- the relation of the two, the *IQ* value of the electron process drive, and
- the intensity of the anti-electron process,
- the most important, however, is that dt_i, the time system of the electron process of all elementary processes is quasi equal and the same; it makes possible for the elementary quantum communication to be free – independently of c_x the quantum speed and *IQ* the intensity of the quantum drive.

The anti-electron process is an important component of elementary communication. This is the tool for controlling the elementary standards of the processes.

Neutron processes are passive and to be driven in elementary communication. But driving (only) the neutron process of the other elementary process of the communication is not enough for completing the elementary cycle. The anti-neutron process and the intensity of the anti-electron process are the ones which ensure that the intensity of the anti-proton collapse, the inflexion and the proton process correspond to the standards of the impacting (communicating) elementary process.

There is no way the anti-process of the impacting elementary process would automatically control the outcome of the process before the impact starts.

the direct line of the impacted process starts by the proton process of quantum speed c_x and intensity ε_x	the impact drives the neutron process of the impacted process and the anti-neutron process is already of quantum speed c_y and intensity ε_y

The challenge of the impacting elementary process is that the first cycle of the process shall be completed. The first impact shall go through. Once the first cycle has happened, the intensity of the anti-neutron process corresponds to the intensity of the neutron process, and the anti-electron process drives the anti-proton collapse in line with the intensity of the impacting communicating elementary process. The infinite number of similar elementary impacts is controlling the continuity of the elementary communication.

The process of elementary communication will be explained in the examples.

Elementary processes are in stable and balanced elementary status. This is the result of the progress of the elementary evolution from the *plasma* state to the *Hydrogen* process. The elementary evolution has also an infinite number of natural isotopes with non-balanced formations. Elementary processes of the periodic table communicate and establish new elementary compositions.

Elementary processes with neutron process dominancy use their increased quantum drive; proton process dominant elementary processes use their electron process *blue shift* surplus.

A.4.1.1. <u>To understand the basics of the elementary quantum communication, first the origin of the water has to be understood. And for understanding the water, the specifics of the *Hydrogen* process have to be accepted.</u>

S.A.
4.1.1

Ref.
S.3.4

The *Hydrogen* process is with infinite low electron process intensity, with infinite low quantum speed and with infinite low intensity of the neutron process – but, most importantly, with infinite low intensities of the anti-neutron and the anti-electron processes as well.

The infinite low speed value of quantum communication characterises the space-time of the *Hydrogen* process. The electron process of the *Hydrogen* process is of infinite low intensity, therefore, it is available for infinite duration and in infinite surplus.

Natural *Oxygen* is with electron process *blue shift* surplus. Because of its reduced quantum speed, the *IQ* quantum drive of the *Oxygen* process is of limited value.

As a result of their quantum communication, the neutrons of the *Hydrogen* and the *Oxygen* processes are driven by the electron process of the *Oxygen*. The infinite low intensity of the electron process of the *Hydrogen* is not capable of driving neutron processes.

As the electron process surplus of the *Oxygen* process has been fully used, for driving the neutrons of the *Oxygen* and the *Hydrogen* processes, the conflict of the electron process *blue shift* surplus of the *Hydrogen* process makes the water naturally liquid. The infinite low intensity of the anti-electron process of the *Hydrogen* does not allow the completion of its first cycle. Therefore, the *Hydrogen* remains being *Hydrogen* process whatever is the drive.

This is the reason why water extinguishes fire: the neutron processes of the *Hydrogen* take all electron process *blue shift* drives of the *Oxygen* process; separately, otherwise feeding the conflict of the fire.

This last statement needs the proof.

The reason for the difference between the impacts of the water and the *hydrocarbons* is the difference in the natural speed values of the quantum communication of the *Carbon* and the *Oxygen* processes.

The natural quantum speed of the *Oxygen* is 299,503 km/sec,

the quantum speed of the *Carbon* process is taken for 300,000 km/sec.

The *blue shift* conflict of the fire is working at the speed value of the quantum communication of the *Earth*. This is: 299,792 km/sec.

With reference to Table 1.1 of Section 1,

the quantum drive of the *Carbon* elementary process is: $IQ = 8.881 \cdot 10^{10} \ km^2/s^2$,

while the quantum drive of the natural *Oxygen* process is only $IQ = 8.847 \cdot 10^{10} \ km^2/s^2$.

[The *IQ* above the *Earth* surface is: $8.987 \cdot 10^{10} \ km^2/s^2$.]

Ref.
Table
1.1
4G1
4G2

In order to be in conflict with the intensity of the electron process of the fire and, in this way, feeding the fire, the water would need to have more *IQ* drive value. Through having more *IQ* drive, water takes away the electron process *blue shift* drive of the *Oxygen* process from the fire. In this way, it is extinguishing fire. (The extinguishing capability obviously also depends on the acting volumes.)

The *hydrocarbon* process with the necessary *additional quantum impact* of the *Oxygen* process (from the natural environment), on the contrary, feeds fire. The *Oxygen* process is not challenging the *Hydrogen* process of the *hydrocarbons*. *Oxygen* process deepens the conflict.

Water cannot impact *hydrocarbon* fire – the *IQ* drive of the *Carbon* process with reference to 4G1 and 4G2 above does not allow the impact from the *Oxygen* process to act.
This is the reason why *hydrocarbon* fire cannot be extinguished by water.

The elementary process structure of the *hydrocarbons* is similar to the water. The *Carbon* process as impact is challenging the neutron process of the *Hydrogen* for completing the first cycle for infinity. The *refractive index* of the water and the *hydrocarbons* well demonstrate this "fight". For the water it is 1.33 and for *hydrocarbons* is around 1.4.
The speed of quantum communication has no impact on the conflict. Conflict is conflict at any speed level, since the conflict is developing at the time system of the electron process. Therefore, *Oxygen* and *Carbon* processes are in conflict and feed fire.

S.A. **A.4.1.2.** <u>The quantum communication of the *Hydrogen Chloride (HCl)* and its acid in</u>
4.1.2   <u>solution with water can also be assessed as a primary example of communication.</u>

There are two "options" for the *Chlorine* elementary process, with higher *IQ* drive than the *Hydrogen* process: driving the neutron process of the *Hydrogen* of less, or driving their own neutron process of higher intensity demand. This, by the way, is the only valid option for all elementary processes, independently of whether they are with electron process *blue shift* surplus or without – but with increased intensity.
As the *IQ* drive of all other elementary process is more than the infinite low value *IQ* drive of the *Hydrogen* process, they will be driving both, but, first of all, the neutron process of the *Hydrogen* process. The "would be anti-neutron process" of the *Hydrogen* process is of infinite low intensity (corresponding to the intensity of the direct neutron process.)

4G3   The infinite low intensity of the electron process of the *Hydrogen* is coming, however, from the infinite length of the neutron process:

$$\lim \varepsilon_H = \frac{\varepsilon_p}{\varepsilon_n} \sqrt{1 - \frac{(c_H - i_H)^2}{c_H^2}} = \frac{\Delta t_n}{\Delta t_p} = \infty$$

With reference to 4G3 on the previous page, the anti-proton process of the *Hydrogen* is of infinite high intensity. There is no elementary process, other than the *plasma* state, whatever is the impact on the neutron collapse of the *Hydrogen* and on the intensity of the anti-electron process, which could produce the drive of this *inflexion*.

Coming back to the *HCl* process, the difference between the drive and the demand is:

$$\Delta_{demand-H} = IQ_X - IQ_{a-H}$$
$$\Delta_{demand-X} = IQ_X - IQ_{a-X} \qquad \Delta_{demand-H} > \Delta_{demand-X}; \quad \text{as} \quad IQ_X - IQ_{a-X} = 0$$

4G4

The neutron collapse of the *Hydrogen* process takes all electron process *blue shift* drives of the *Chlorine* process. At the same time, the *IQ* drive of the *Hydrogen* process (as being of infinite low intensity) is not capable of driving the neutron processes of the *Chlorine*.
This is the reason why *Hydrogen Chloride (HCl)* is a strong and corrosive mineral:
- the *Chlorine* process is missing the part of its neutron process drive and cover;
- as a consequence, the neutron process of the *Chlorine* takes the *IQ* quantum drive and the proton process cover of other elementary processes.

In the case of the solution of *HCl* with water, the case is even worse. The *Hydrogen* process of the water deepens the balance problem. The water is liquid because of the electron process *blue shift* surplus and conflict of the *Hydrogen* process. The *Hydrogen* process of the water takes even more *Chlorine* process *IQ* drives, resulting in increased electron process *blue shift* drive and proton process cover demand. The consequence is the highly pungent solution. More water dilutes the solution as the demand increases while the

Chlorine process source remains unchanged. More *HCl* on the contrary solidifies the solution: as the relative demand of the *blue shift* surplus of the *Hydrogen* process is decreasing.

The same is the logic of the quantum communication of naturally gastric acids, the solution of *Potassium Chloride (KCl)* and *Sodium Chloride (NaCl)* with water, *Sulphuric* acid and others. *Hydrogen* and the *Oxygen* processes take the natural quantum drives of elementary processes.

A.4.1.3. <u>Communication of elementary processes with proton and neutron process dominance in general</u>

The basis of the quantum communication is the intensity of the anti-electron process and the intensity of the neutron process.

The general rule of elementary communication is: electron processes drive neutron processes with less intensity. Low intensities invite electron process drives. Higher *IQ* potentials, with reference to 4G4, first of all drive neutron processes with the lowest intensity demand. The intensity demand has been established by the intensity of the anti-neutron process. The anti-neutron process belongs to the neutron process of the drive and the anti-proton process is driven by the anti-electron processes of the driving elementary process as well. The remaining elementary relations follow the same rule.

Hydrogen process and water behave differently, as it has been presented in the previous sub-sections.

Diag. 4.6 demonstrates the case, establishing cross-relations between the proton and the neutron processes of the elementary processes. The priorities of the relations depend on the intensities of the electron process *blue shift* drives.

A Proton process dominance and electron process *blue shift* surplus	proton process *IQ* is of decreased value but with surplus *acting as consequence* neutron process	anti-proton process ↑ anti-neutron process *anti-IQ* is of increased value	→ proton process A
B Neutron process dominance and increased electron process intensity	*first acting* proton process *IQ* is of increased value neutron process	anti-proton process ↑ anti-neutron process *anti-IQ* is of decreased value and with surplus	B → proton process

Diag.4.6

There is a simple elementary communication of two elementary processes A and B presented in Diag.4.6. A is proton process dominant with electron process *blue shift* surplus. B is neutron process dominant and with increased electron process intensity. The cross drives of the neutron processes ensure the communication.

There could be an average quantum speed value calculated for the operating communication. This speed value would be between the speed values of the elementary processes, proportional to the intensity relations.

With reference to Section 2, the balance will be at the equality of the quantum drives:

4H1
$$c_A^2 \cdot \varepsilon_A = c_B^2 \cdot \varepsilon_B \quad \text{and} \quad n\frac{c_A^2}{\varepsilon_A} = \frac{c_B^2}{\varepsilon_B}$$

S.A.
4.1.4

A.4.1.4. <u>In the case of quantum communication between elementary processes, both with proton process dominance, the quantum speed value is important. There is an electron process *blue shift* surplus on the direct sides of both elementary processes.</u>

The elementary process with higher quantum speed than $c_{Earth} = 299{,}792$ km/sec of the *Earth* surface drives the anti-proton process to full collapse with increased intensity.

The increased intensity is the consequence of the direct process. The increased collapse of the anti-proton process is necessary, since the elementary process is proton process dominant.

The gaseous status would mean electron process *blue shift* conflict with the *Quantum Membrane* of the *Earth* surface.

But there is no conflict as the quantum speed is of increased value. Therefore, while the direct process is with electron process *blue shift* surplus, it has solid status.

4I1

It seems obvious,		having $\varepsilon_x > 1$ on the direct side
	since $\varepsilon_{-xm} = \dfrac{1}{\varepsilon_x}$	means $\varepsilon_{-xm} < 1$ on the anti-process side.

Elementary process with natural quantum speed of the elementary evolution less than the quantum speed of the *Earth* surface is speeded up by the quantum drive of the *Earth* to $c_x = 299{,}792$ km/sec. The increased by *gravitation* intensity of the electron process generates conflict on the direct side and, therefore, the aggregate status of these elementary processes is gaseous.

The natural elementary communication, the natural elementary mix, the result of elementary evolution of these two elementary processes, results in solid aggregate state with electron process *blue shift* surplus. The average quantum speed of the mix is a speed value between the two processes.

The elementary communication is difficult if the difference between the quantum drives of the elementary processes is of increased value.

S.A.
4.1.5

A.4.1.5. <u>Elementary communication of *acids* and *bases*</u>

The intensity of the quantum drive is the decisive component.

Elementary process with certain quantum drive intensity cannot impact other elementary processes with higher quantum drive, but can invite those for communication. In line with Diag. 4.6, the elementary process with quantum drive of higher intensity is the one starting the communication, while the elementary process with less electron process intensity is "offering" its neutron process to be driven.

The difference between the impacts of acids and bases is the way they communicate.

In the case of acids, like H_2SO_4, HNO_3, H_2PO_3, H_2BO_3 and other inorganic acids, the elementary communication is quasi completed.

If acids are impacting elementary processes or a mix of elementary processes, the impacted processes communicate with all elementary processes of the acids, but first of all with the *Hydrogen* process, as this is the one with infinite low intensity drive and always in demand. (*Hydrogen* takes infinite number of electron process *blue shift* impact and proton cover potential from elementary processes – without any compensation.)

The *Sulphur, Nitrogen, Phosphorus, Boron* and the *Oxygen* processes of the above acid examples have already lost their electron process intensity drive and proton process cover potential to the *Hydrogen* process. (Losing on the proton process potential means that while the electron process *blue shift* impact is used as drive, the proton process cover would only be taken if the neutron collapse had been completed.)

Therefore, they will be acting as inviting the impacted elementary processes in order to utilise their quantum drive. As a benefit of the communication, the missing electron process intensity and proton process cover will be given by the impacted elementary processes. In this way, the impacted elementary process or mix will be destroyed.

This damage is not about the destruction of the relation between the intensities of the neutron and proton processes, which is the characteristic of isotopes. This damage causes missing intensities in the impacted elementary processes. *Sulphur, Phosphorus and Boron,* as shown above, have increased impact, *Oxygen* and *Nitrogen* have less.

The elementary process of *bases* like *KOH, NaOH, Ca(OH)$_2$* and others is quasi open.

The communication gives similar results as in the case with acids: destruction, but bases dissolve the impacted process.

The reason in conventional terms is the *(OH)*.

At the start of the communication with the elementary processes of the base, like *Potassium, Sodium, Calcium*, the electron process quantum drive and proton process potential of the *Oxygen* process has already been utilised by the *Hydrogen* process.

The internal elementary communication of the base still goes on when the elementary process or processes are impacted. The elementary communication with the *Ka, Na, Ca* processes becomes expanded, including the impacted elementary processes as well.

A.4.1.6. <u>Quantum communication without the physical contact of the elementary processes. Quantum impulse from quantum membrane to quantum membrane without direct contact, via the separating boundaries.</u>

S.A.
4.1.6

We have to distinguish elementary communication from the quantum impacts of systems. Elementary communication, the proton/neutron–anti-proton/anti-neutron process relations are reciprocal. In the case of quantum systems, the one of increased intensity impacts the other with less intensity value.

Quantum systems are characterised by the same parameters as elementary processes: the quantum speed and the intensity of the quantum membrane. Having two systems x and y and looking for their relationship, the one with higher *IQ* value has quantum impact over the other. The quantum speed, in this case, depends on the *blue shift* conflict within the *Quantum Membrane* of the system.

$$\frac{dmc_x^2}{dt_i\varepsilon_x}\left(1-\sqrt{1-\frac{(c_x-i_x)^2}{c_x^2}}\right); \quad \text{and} \quad \frac{dmc_y^2}{dt_i\varepsilon_y}\left(1-\sqrt{1-\frac{(c_y-i_y)^2}{c_y^2}}\right); \qquad \text{4J1}$$

In elementary relations if $c_x > c_y$ and as consequence of the

$c\cdot\varepsilon = const$ rule, $\varepsilon_x < \varepsilon_y$ the relation of the quantum drives is obvious: $\quad \dfrac{c_x^2}{\varepsilon_x} > \dfrac{c_y^2}{\varepsilon_y} \qquad$ 4J2

4J3   In the case of quantum systems, the *IQ* value can be consequence of the conflict of the quantum membrane for which: $\dfrac{c_x^2}{\varepsilon_x} \le n\dfrac{c_y^2}{\varepsilon_y}$

4J4   and $\dfrac{dmc_x^2}{dt_i\varepsilon_x}\left(1-\sqrt{1-\dfrac{(c_x-i_x)^2}{c_x^2}}\right) \le n\dfrac{dmc_y^2}{dt_i\varepsilon_y}\left(1-\sqrt{1-\dfrac{(c_y-i_y)^2}{c_y^2}}\right)$

The quantum membrane with higher *IQ* quantum drive intensity is the one impacting. Quantum membrane with quantum drive of less intensity cannot have impact on a quantum system of higher intensity.

Higher quantum drive means either elementary process with higher *IQ* drive (higher electron process intensity and higher quantum speed) or *IQ* drives of less intensity but in increased in numbers.

It is difficult to increase the intensity of a system with quasi open boundaries.

The *IQ* value of the *Quantum Membrane* of the *Earth* is quasi constant and it is:

4J5 $8.99\text{E}+10\,\dfrac{km^2}{sec^2} = 8.99\text{E}+16\,\dfrac{m^2}{sec^2}$

This means the *IQ* drive of the *Earth* surface remains quasi constant.

4J6
4J7   {The *Planck* constant is $h = 6.626,069,572\,\text{E-}34\,\text{J}\cdot\text{sec}$,

but $E = h\gamma$ is about the absolute energy of a single quantum impact, while the electron process *blue shift* impact values in 4J1 and 4J4 are about intensities.}

Generating a quantum impact within the quantum system of the *Earth* with constant *IQ* quantum drive means impacting the quantum system of the *Earth*. The impact generates a signal with certain intensity and this intensity is transferred by the system. The frequency of the impact is the sequence of the effect.

The *Earth* system has its intensity characteristics and this is impacted. The external impact is periodic in certain frequency. The description is:

4K1 $\gamma_X\dfrac{dmc_E^2}{dt_i\varepsilon_E}\left(1-\sqrt{1-\dfrac{(c_x-i_x)^2}{c_x^2}}\right);$ or $=\dfrac{dmc_E^2}{dt_i\varepsilon_E\Delta t_X}\left(1-\sqrt{1-\dfrac{(c_x-i_x)^2}{c_x^2}}\right)$

The impact is propagating by c_E the quantum speed of the system.

4K2   The certain periodic impact in time determines the length of a single impact: $\lambda_X\cdot\Delta t_X = c_E;$ or $\dfrac{\lambda_X}{\gamma_X}=c_E$

4K3   With reference to 4J4 the impact is increased. $\gamma_X > 1$ and the acting intensity is: $\varepsilon_X = \dfrac{\varepsilon_E}{\gamma_X}$

Taking the systems in 4J1 *x* and *y* the quantum impact and the relations of the systems can be assessed.

4K4

$dt_X = \dfrac{dt_o}{\sqrt{1-\dfrac{v^2}{c_x^2}}};$	$\varepsilon_X = \dfrac{1}{dt_X}$ $\varepsilon_x \neq \varepsilon_X$	$dt_Y = \dfrac{dt_o}{\sqrt{1-\dfrac{v^2}{c_y^2}}};$	$\varepsilon_Y = \dfrac{1}{dt_Y}$ $\varepsilon_y \neq \varepsilon_Y$	The impact is presented by *v* and the system is presented by *c* the speed of quantum communication

If $c_x > c_y$ it follows that $dt_y > dt_x$. The time count in system *y* is more than in system *x*.

With reference to the *Planck constant* in 4J6 and the *Einstein-Planck equation* in 4J7, the energy of a *single* quantum impact is equal to

$$1 \cdot \frac{dmc_E^2}{dt_i \varepsilon_E} \left(1 - \sqrt{1 - \frac{(c_x - i_x)^2}{c_x^2}} \right) = \gamma \cdot h \qquad \text{4L1}$$

from which follows [taking the mass impact of the electron (process) for $\Delta m = 9.10938215 \times 10^{-31}$ kg] that

$$\frac{1}{\Delta t_i \varepsilon_E} \left(1 - \sqrt{1 - \frac{(c_x - i_x)^2}{c_x^2}} \right) = 8.093 \times 10^{-21} \ 1/\text{sec} \qquad \text{4L2}$$

+

The lesson is that our quantum system on the *Earth* surface can be impacted by increased quantum signal, and the signal can be received by equipment properly tuned to the frequency of the impact.

4.5.7. Anti-processes, the basis of quantum communication

With reference to the equality of $c_x^2 \cdot \varepsilon_x = c_y^2 \cdot \varepsilon_y = ... = c_n^2 \cdot \varepsilon_n = const$
the intensity of the anti-process quantum drives are also equal, as $\quad \varepsilon_x = \dfrac{1}{\varepsilon_{-x}}$

4M1

$$\text{This way:} \quad \frac{c_x^2}{\varepsilon_{-x}} = \frac{c_y^2}{\varepsilon_{-y}} = ... = \frac{c_n^2}{\varepsilon_{-n}} \qquad \text{4M2}$$

4M2 above means, all anti-process quantum drives are equal!

This makes the quantum communication not just easy, but this is its most important factor. In the case of the communication of two or more elementary processes, the quantum speed and the intensity coefficient of the acting elementary side will be modified in line with the required anti-drive of the anti-proton process of the other communicating elementary process. The intensity coefficient is in linear function of the quadrat of the quantum speed.

The anti-process drive of each elementary process of the periodic table makes it possible to communicate with any of the other elementary processes.

The equality of the anti-process drives also means: the surplus in anti-electron process *blue shift* impact of elementary processes is established by the direct process.
The higher the intensity of the quantum drive of the direct process is, the higher is the anti-electron process *blue shift* surplus.

$IQ_x = \dfrac{c_x^2}{\varepsilon_x}$; and $\varepsilon_{-x} = \dfrac{1}{\varepsilon_x}$; $\quad$ the lower is the value of ε_x,
the higher is the difference in the absolute values $\quad \varepsilon_{-x} > \varepsilon_x \qquad$ 4M3

The higher the intensity of the direct quantum drive is, the less is the need in anti-electron process *blue shift* impact, since the demand of the anti-proton process in *blue shift* drive is less. The surplus can be used for loading gravitation or other purposes.
In the case of proton process dominant elementary processes, the anti-process quantum drive is the one being in overload relative to the direct drive: $\quad \varepsilon_{-x} < \varepsilon_x \qquad$ 4M4

Therefore, the process has electron process *blue shift* surplus for initiating quantum communication.

**Annex
5.1**

Rotating disc experiment with switching off and on light impacts
(Explanation of the experiment, made in 2009.)
[*Youtube*: https://www.youtube.com/watch?v=xXCGmH_k_n0]

The rotating disc experiment proves, the lights are off and on, depending on the frequency of the rotation of the disc.

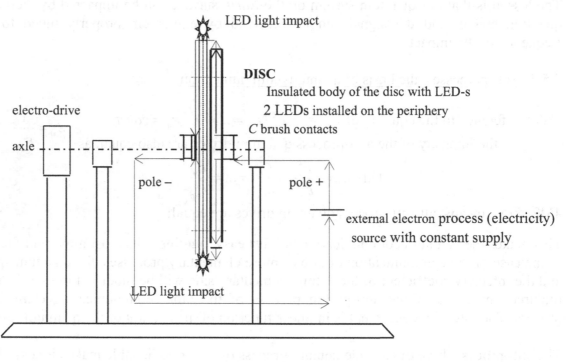

**Fig.
5.3**

Fig.5.3

Rotation (acceleration) is increasing the intensity of the elementary process.
But the intensity of the electron process is stable, since this guaranties the elementary process itself.

5F1
$$(x \cdot \varepsilon_n) - (x \cdot \varepsilon_p) = x \cdot (\varepsilon_n - \varepsilon_p); \quad \text{while} \quad \varepsilon_x = \frac{\varepsilon_p}{\varepsilon_n} = \frac{x\varepsilon_p}{x\varepsilon_n} ;$$

With the increase of the acceleration (the intensity), the gap between the physical impact (the acting numbers for the unit period of time) of the proton and neutron processes is changing.

    In the case of proton process dominancy the intensity increase automatically results in increased number of electron processes, additional electron process *blue shift* surplus and conflict.

    In the case of neutron process dominance the difference in the acting absolute numbers of the neutron and the proton processes is growing, while the relation remains the same.

But the neutron collapse needs drive! There is no collapse without drive.

The intensity of the electron process is sufficient for the collapse, but under the increasing accelerating impact of the rotation, the mechanical effect of the acceleration creates an increased demand:

- the increasing acceleration results in the increasing intensity of the elementary process; but the intensity of the electron process remains unchanged;
- the relation of the intensities of the proton and neutron processes is the same, but the demand in electron process *blue shift* drive in numbers is increasing, while the supply remains behind - which causes de facto electron process intensity deficit.

  [The intensity of the electron process remains the same, and covering the elementary demand the only way is the increase in numbers.)

In the case of the continuous increase of the acceleration (increase of the frequency of the rotation), this functioning problem is growing and growing. With remaining at constant acceleration (at constant speed of the rotation) in the case of a homogeneous disc the functioning is consolidating. The elementary balance between the proton and neutron processes is stabilising: the electron process supply by the proton process is catching up with the neutron process demand. In the slowing down status, the increased to the accelerating conditions electron process *blue shift* surplus is given off by the elementary process, as the neutron process demand is decreasing.

In the case of an electron process supply from an external source of limited value, the impact is changing: If there was a light impact conflict without the rotation of the disc (acceleration), in higher frequency the conflict disappears, as the conflicting in normal conditions electron process number, as result of the increased demand disappears. With the slowdown of the rotation (frequency) the light impact is acting again, as the conflict is recreating.

The mechanism of the conflicting period is, that

- the anti-neutron process in fact is "sensing" a decreased intensity relative to the necessary elementary demand (the less in numbers electron process *blue shift* drive even of the standard to the elementary process intensity means less integrated impact);
- for keeping the elementary balance, the generating anti-electron process *blue shift* impact, which otherwise is with certain surplus, now will be using part of the surplus in order to collapse the anti-proton process by the necessary to the increase of the elementary process intensity;
- the missing in numbers electron processes on the direct side results in missing anti-electron process *blue shift* surplus on the other side;
- there is no difference in the resulting impact between the electron and the anti-electron processes;
- the electron process intensity demand is formulating in fact on the anti-direction side, where the cover is not necessary, as the surplus is the one which is missing!!
- the elementary process in functioning problem takes anti-electron process *blue shift* impact from aside wherever it can be taken from (all anti-electron processes are of equal *IQ* drive);
- if the accelerating impact is so serious and the gradient of the increase of the acceleration is so high that the elementary process cannot manage the change – it falls apart, the rotating system collapses with mechanical damage;
- the higher the Periodic Number is, the higher is the anti-electron process *blue shift* surplus reserve of the elementary process;
- the closer the elementary process to the $\varepsilon_x = 1$ status is, the easier is to the proton process to provide the necessary number of the electron processes.

If a rotating disc has been left alone rotating, the internal distribution of the anti-electron process balance will be the source for covering the missing electron and anti-electron process intensities – until the mechanical structure of the disc is capable to do so without destroy.

The case is relativistic.

The change of the speed of the rotation is modifying the circumstances of the elementary processes of the wires of the halogen lights. The light impact means conflict.

The electron processes from the external source are in conflict with the electron processes of the elementary process of the wires of the halogen lightbulbs.

The speed difference, caused by the rotation is minor-minor relative to the speed value of the quantum communication, but the intensity of the conflict is changing. The increase of the speed of the rotation intensifies the elementary process.

The intensity of the electron process is without change, the relation between the intensities of the proton and the neutron processes remains as it was, but the number of the developing and utilised electron processes - for the unit period of time - becomes increased = the elementary process is of an increased tempo.

5F2
$$\varepsilon_{ex} = \frac{\varepsilon_p}{\varepsilon_n}\sqrt{1-\frac{(c_x-i_x)^2}{c_x^2}}; \quad \text{and} \quad \Delta n \frac{dmc_x^2}{dt_i\varepsilon_{xe}}\left(1-\sqrt{1-\frac{(c_x-i_x)}{c_x^2}}\right) = \frac{dmc_x^2}{dt_i\frac{e_{ex}}{\Delta n}}\left(1-\sqrt{1-\frac{(c_x-i_x)}{c_x^2}}\right);$$

$$\Delta n = n_b - n_a$$

The format in 5F2 is equivalent to an elementary process with increased neutron process intensity, this way obviously with increased electron process demand.

While the intensity of the external electron process supply remains unchanged, the utilisation of the internal electron process becomes increased. As consequence the conflict becomes reduced.

If we suppose that the number of elementary cycles in normal elementary conditions of the wires is n_a and this number is changing to n_b, as consequence of the rotation, the difference is the one causing the impact.

The change in the numbers, the intensity increase is direct consequence of the modified time flow:

5F3

$dt_b = \dfrac{dt_a(=1)}{\sqrt{1-\dfrac{v^2}{c_x^2}}};$	as $v = r \cdot \omega = r \cdot f$ and f is the frequency of the rotation	$\Delta n \approx \Delta t$ the increase in numbers is proportional to the time increase

dt_b is now the one corresponding to the unit period of time of the elementary process.

For having the elementary process unchanged, with the standard period increased.

5F4   The standard electron process number relation of the elementary process is: $\dfrac{dn_{st}}{dt_{st}}$;

5F5   If now $dt_{st} = dt_b$ (as the elementary process did not change): $dn_{st} = dn_b$

The increase of the denominator of ε_{ex} in 5F2 results in the increase of the intensity.

The other format of the description of the acceleration with reference to 5F2 could be:

5F6
5F7

$e_{\Delta x} = \dfrac{dmc_x^2}{dt_i\varepsilon_{ex}\sqrt{1-\dfrac{v^2}{c^2}}}\left(1-\sqrt{1-\dfrac{(c_x-i_x)}{c_x^2}}\right);$	And from 5F6: $e_{\Delta x} = e_{x+} - e_x$ and $e_{x+} = n_{x+}e_x - n_x e_x$

Whatever is the scale of the change, the change is change and the consequences are consequences in line with the intensities of the space-times.

In a case of a change, caused by the 6000 rpm rotation, the modification of the conflict of the electron processes within the halogen lights is at the level of *minus twelve* after the decimal point: $\Delta n = 5.563E-12$.

5F8

The intensity of the change within the elementary process as above, the appearance of this change within our space-time on the surface of the *Earth* results in clear event-consequence: switching off and on lights.

The result in 5F8 relates to our space-time, with our time count is in our time format.

With reference to the *pyramid* experiments, the anti-electron process voltage supply from the *Earth* is the impact of *gravitation* indeed.

The lesson is that the increase in numbers of the electron processes means increased voltage potential:

- the increasing number of the missing electron process *blue shift* impacts causes voltage increase in the rotating disc experiment;
- the *Aluminium* process contact to the *Earth* in the pyramid experiments has increased with its different elementary structure the voltage potential.

A *pyramid* with increased geometry will be generating increased voltage.

**Annex
6.1**

***Balanced Concrete* structure (*Balance-mix*)
for compensating the impact of *water veins* and *Hartmann* and other type of *grids***

**SA
6.1.1**   A.6.1.1. Impact of *water veins*:

Water veins do not directly impacting us.
The damaging impacts to our balanced health system are coming from the unbalanced mineral compositions within the *Earth* below us. Impacts become strengthened by the water flow in the veins.
The acting electron process *blue shift* surplus of *water veins* is impacting the elementary processes of minerals. The electron process *blue shift* surplus of the water intensifies the quantum communication. The intensity increase may also cause *blue shift* conflict.
The reason is the *Hydrogen* process.
The infinite low intensity of the neutron process of the *Hydrogen* process, its quasi never ending character is the one which impacting other elementary processes: "inviting" the electron process *blue shift* impact of the elementary processes of minerals to communicate.

Elementary processes of the soil close to the *Earth* surface and sometimes also minerals deeper inside are with close to balanced status, independently are they with proton or neutron process dominance. *Water veins* bring electron process *blue shift* surplus in and the permanent water flow ensures the continuity of the *blue shift* supply.

The impact is twofold:

> If the region around the *water vein* is with soil and/or with minerals of *proton process dominant* elementary processes, the *Hydrogen* process portion of the water takes their *blue shift* surplus and the proton process cover. This way weakens the energy intensity potential (the available electron process *blue shift* surplus and proton process cover) of the region.
> The *Hydrogen* process takes all the available electron process *blue shift* impact of the natural surplus of the *Oxygen*, still having capacity to engage more, as its natural *IQ* value is of infinite low value.

6M1 $$\lim IQ = \lim \frac{c_H^2}{\varepsilon_H} = 0 \quad \text{as} \quad \begin{array}{l} \lim c_H = 0 \text{ and} \\ \lim \varepsilon_H = \infty \text{ (meaning infinite low intensity)} \end{array}$$

The *IQ* value of the *Hydrogen* process remains of infinite low value even within the water, with quantum communication at the quantum speed of the *water*, significantly less than $c = 299{,}792$ km/sec the quantum speed of the *Earth*.
The infinite low *IQ* drive means infinite high capacity for the neutron processes to be driven.
Hydrogen is capable for taking all *blue shift* impacts. This is the reason *water veins* contain elementary processes as of dissolved minerals. Mineral waters, having increased, already conflicting impacts from minerals may also have increased temperature, as the indicator of their internal electron process *blue shift* conflict.

> If the region of the *water vein* is with minerals of <u>*neutron process dominant*</u> elementary processes, the acting electron process *blue shift* surplus of the water flow is impacting the elementary communication of the region: intensifies the electron process *blue shift* drive of the neutron processes of the minerals.

$$\frac{dmc^2_{water}}{dt_i \varepsilon_{water}}\left(1-\sqrt{1-\frac{(c_w-i_w)^2}{c_w^2}}\right) + \frac{dmc^2_{min}}{dt_i \varepsilon_{min}}\left(1-\sqrt{1-\frac{(c_m-i_m)^2}{c^2}}\right) > \frac{dmc^2_{min}}{dt_i \varepsilon_{min}}\left(1-\sqrt{1-\frac{(c_m-i_m)^2}{c^2}}\right)$$ 6N1

The direct consequence is demand in electron process *blue shift* drive and proton process cover from external source, as the mineral (alone at the region) cannot provide the drive and the proton process cover at this increased intensity.

ε_{min}, the intensity of the mineral before the impact becomes: $x \cdot \varepsilon_{min} = \varepsilon_{imp}$, after it.   6N2

ε_{imp} means different elementary relation or relations.

For re-establishing the balance the region needs external electron process *blue shift* drive and proton process cover support for managing the increase, as water cannot provide it. This way the water flow deepens the energy need.

The <u>*Hydrogen*</u> process is with electron process *blue shift* surplus. While its neutron process is of infinite low intensity, the speed value of its quantum communication corresponds to the quantum speed of the *water*.

[The available electron process *blue shift* surplus of the *Hydrogen* process and the conflict – caused by the infinite low intensity of the neutron process – makes water liquid; if heated, the conflict is increasing and water is steaming].

The acting electron process *blue shift* surplus of the water, with reference to 6N1 increases the intensity as result of the impact. The region tries to provide the missing proton process cover.

While the first option with <u>*proton process dominant elementary processes*</u> seems to be with less damaging external impact, the second with <u>*neutron process dominancy*</u> seems to be with more. The "action of the nature" in both cases is of the same principle: *re-establishing the balance*!

The external consequences in both cases are similar: loss of the energy/mass potential of the soil. Elements and minerals of the soil need external energy source. Living organisms and human systems can be the subjects and the source for covering the missing electron process *blue shift* impact and proton process potential.

Nature is resolving the balance deviation.

For re-establishing the balance the elementary processes of the soil are taking electron process *blue shift* impact and proton process cover from all available sources above the *Earth* surface, including the human systems.

Soil close to the surface is with minerals of life important elementary processes. The intensities of the proton and the neutron processes are close to equal. In the other direction in the depth of the *Earth* elementary systems are with neutron process dominant elements.

As consequence, human elementary systems may temporarily lose life important elementary components and may have balance problem. Losing balance is the one with more damaging impact.

A.6.1.2. Impact of the *Hartmann* grid, knots and lines:

The quantum system around the *Earth* is established by the quantum impulse of elementary processes. The quantum impact of *gravitation* (the anti-electron process *blue shift* impact of elementary processes) results in electromagnetic features.

The approaching function of *gravitation*, the mechanical impact of the sphere symmetrical expanding acceleration causes *blue shift* impact and also generates the *Quantum Membrane* above the *Earth* surface.
The quantum impact of the *Sun* has its *blue shift* impact as well.
The conflict between the electron process *blue shift* impact from the *Sun* and the *blue shift* impact of *Earth gravitation* rotates the *Earth* around its own *North-South* axle. The two *blue shift* impacts cause rotation as this is the easiest way for resolving the conflict. [The conflict can also be characterised by the temperature on the surface of the *Earth* as additional indicator.]

The rotation of the *Earth* and the quantum impact of *gravitation* (the quantum impact of the anti-electron processes of elementary process through the *Earth* surface) generate magnetic features above the *Earth* surface. The case is similar to the experienced magnetic effect at the two ends of the iron core of electromagnets; just in that case electricity is the one flowing/circulating within the wires of the electromagnet.

The moving elementary proton/neutron relation, the reason of magnetic phenomenon is closing externally through the *North* and *South* poles. Magnetic impact is measured on the surface of the *Earth*. In addition the axle of the rotation of the *Earth* is declined relative to the impact from the *Sun*, further contributing to the development of the magnetic effect.

We live within this quantum system *space-time*.
We are also generating quantum impact and quantum impulse; we are also the participants of the elementary communication and processes; we are the beneficiaries of and also the ones suffering from the quantum communication of the system.
This is also the *Quantum Membrane*, with the quantum impacts of *water veins* and all other elementary processes, where all balance deviations have been managed by the nature; with quantum energy intensities taken from one side and given to others as the balance deems it.

The *blue shift* impact of the *Sun* to the *Quantum Membrane* of the *Earth* is not constant. The impact during the day (from morning to evening) is the "yes" category, the night impact is category "quasi no".
The intensity and the balanced status of the *Quantum Membrane* around the *Earth* therefore is different under "yes" and under "quasi no" periods. (Sensitive systems do even feel the difference of the load and the impact between sunny and clouded days.)
The *Quantum Membrane* we live on the *Earth* surface in does communicate and compensate these differences for finding always the actual balance. Compensation works in two directions as for the case with the *water veins*:
The quantum impact demand of the *Quantum Membrane*
  ➢ during the "quasi no" night periods may be increased, deepening this way our quantum energy losses; or
  ➢ establishing the balance or giving us plus energy during "yes" periods,
but resulting changes of the balanced status in both cases.

Especially serious is the case if we are losing on our personal *blue shift* impact and proton process cover to the *Quantum Membrane* during the daily period. (As the load during the night without the *Sun* impact is anyway more.) This is the case if *water vein* problems become escalated by *Hartmann grid* and other impacts.

A.6.1.3. The solution:

Protecting, shielding structures, constructions built on the *Earth* surface should themselves be in integrated elementary balanced status. This type of construction keeps back the damaging quantum impacts and does not influence the elementary processes of those living above.

The shield is special concrete structure, mix of different minerals with life important elements with close to balance elementary status, called: the *Balance-mix*.
The protecting concrete shield structure is built in layers with the use of special technology.

**Annex
6.2**

Process based assessment of the "walkers" of oil bath
Note to the articles on "Classical quantum" (11 November 2013, Physics World)
"Spin revolution" (11 January 2014, New Scientist)

Oil is *Hydrocarbon*.

Independently on the exact formula of its composition, the principle of the elementary communication of *Hydrocarbons* is as follows: the electron process *blue shift* of the *Carbon* process drives the neutron processes of the *Hydrogen*, which otherwise driven by the electron process of the *Hydrogen* process of infinite low intensity.

6O1 $\quad e_C = \dfrac{dmc_C^2}{dt_i \varepsilon_C}\left(1 - \sqrt{1 - \dfrac{(c_C - i_C)^2}{c_C^2}}\right)$	c_C and ε_C are the speed of the quantum communication and the intensity coefficient of the *Carbon* process; $i_C = \lim a\Delta t = c_C$ - the acting speed value of the sphere
6O2 $\quad dt_i = \dfrac{dt_o}{\sqrt{1 - \dfrac{i_C^2}{c_C^2}}} \cong \dfrac{dt_o}{\sqrt{1 - \dfrac{i_x^2}{c_x^2}}} \cdots$	symmetrical acceleration of the electron process; dt_i - represents the time system of the electron process.
x denotes any other element	Electron processes of elements are communication.

Carbon is the closest elementary process to the equal proton-neutron process balance.

The mass/energy balance of the elementary processes in absolute terms is:

$$6O3 \qquad \frac{dmc^2}{dt_p \varepsilon_p}\left(1 - \sqrt{1 - \frac{i^2}{c^2}}\right) = \frac{dmc^2}{dt_n \varepsilon_n}\xi\sqrt{1 - \frac{(c-i)^2}{c^2}}\left(1 - \sqrt{1 - \frac{i^2}{c^2}}\right)$$

$$6O4 \qquad \varepsilon_e = \frac{\varepsilon_p}{\varepsilon_n}\sqrt{1 - \frac{(c-i)^2}{c^2}} \qquad \text{(no dimension for the electron process intensity coefficient}$$

6O5   indeed.)

In intensity terms: $\dfrac{dmc^2}{dt_p}\left(1 - \sqrt{1 - \dfrac{i^2}{c^2}}\right) \neq \dfrac{dmc^2}{dt_n}\xi\sqrt{1 - \dfrac{(c-i)^2}{c^2}}\left(1 - \sqrt{1 - \dfrac{i^2}{c^2}}\right)$

6O6   For the *Hydrogen* process:

$\lim \varepsilon_n^H = 0$; and $\lim dt_n^H = \infty$; and the intensity coefficient of the electron process of the

Hydrogen is: $\lim \varepsilon_e^H = \infty$ (reciprocal value to the intensity)

This way all free (but limited number) of electron process *blue shift* drives of the *Carbon* process will be driving neutron processes of the *Hydrogen* process; while the electron processes of the *Hydrogen* processes remaining in surplus are resulting in *blue shift* conflict – reason of the liquid state.

(The increased conflict may result in gaseous state. If the electron process *blue shift* surplus is taken away – by cooling, the *Hydrocarbon* compositions can also be turned into solid state.)

Carbon is elementary process with speed of quantum communication higher value than the quantum speed of the *Quantum Membrane* above the *Earth* surface; with slight proton process dominance against the neutron process, with the intensity of the electron process slightly above 1: $\varepsilon_C > 1$.

Vibration increases the intensity of the internal electron process and the *blue shift* conflict of the *Hydrogen-Carbon* elementary communication of the oil bath.

If the intensity of the *blue shift* impact of the elementary communication without vibration is:

$$e_x = \frac{dmc_x^2}{dt_i \varepsilon_x}\left(1 - \sqrt{1 - \frac{(c_x - i_x)^2}{c_x^2}}\right) \qquad \text{6P1}$$

x in the index means *Hydrocarbon*

Vibration increases the intensity of the electron process.

The *blue shift* impact is about the fully used electron processes of the *Carbon* element and the increased intensity of the *blue shift* surplus of the *Hydrogen* process.

with vibration on, it will be:

where $\sqrt{1 - (v^2/c^2)}$ in the denominator represents the increased intensity, result of the speeding up by vibration

$$e_{xv} = \frac{dmc_x^2}{dt_i \varepsilon_x \sqrt{1 - \frac{v^2}{c^2}}}\left(1 - \sqrt{1 - \frac{(c_x - i_x)^2}{c_x^2}}\right) \qquad \text{6P2}$$

Vibration is specific motion – a speeding up for short length and with inflexions.

The impact is even more increased, since the vibration effect at the inflexions is always doubled and equal to: $[e_x - (-e_x)]$.

This way the elementary processes of the oil bath are speeded up and the electron processes are of increased intensity.

Any oil drops released on the oil bath surface are of normal (as per 6P1) *Hydrocarbons* electron process intensity and surplus, therefore

1. Drops cannot be taken into the bath (as surplus does not take surplus);
2. The impacting by vibration electron process *blue shift* surplus of the bath of increased intensity is *blue shifted* and reflected back at the surface, from the oil drops "walkers", which are also with *blue shift* surplus, just of less integrated intensity.
   The elementary systems with *blue shift* surplus are in conflict.
3. The conflict difference between the intensities of the electron processes of the oil bath (6P2) and the oil drops "walkers" (6P1) results in waves and all those "walkers" symptoms written in the article in the title.

The increased electron process *blue shift* impact can also be written in frequency terms:

$$e_{xv} = \frac{dmc_x^2}{dt_i \varepsilon_x \sqrt{1 - \frac{v^2}{c^2}}}\left(1 - \sqrt{1 - \frac{(c_x - i_x)^2}{c_x^2}}\right) = n\frac{dmc_x^2}{dt_i \varepsilon_x}\left(1 - \sqrt{1 - \frac{(c_x - i_x)^2}{c_x^2}}\right)$$

n corresponds to increased number of electron processes

This is in fact the right formulation, as the intensity of the electron process of the *Hydrogen* process always remains of infinite low value!

The number of the cycles of the vibration for unit period of time (frequency) is equivalent to the motion with speed *v*. The purpose of the vibration is speed increase: having the *Hydrocarbon* with its quasi balanced *Carbon* process and quasi open (as of infinite length) *Hydrogen* cycle in motion. Coming through inflexions is equivalent to speed increase. The inflexion in this case is clearly a mechanical one, result of the vibration (– this is not about the inflexion of the elementary process.)

Speed increase in the denominator in 6R1 means intensity increase.	as $dt_i\sqrt{1 - \frac{v^2}{c^2}} = dt_x$ $dt_x < dt_i$	meaning: the same process happens for shorter period of time.

The frequency value is function of the acceleration and can be expressed as:

6R3 $n = \dfrac{1}{\sqrt{1-\dfrac{v^2}{c^2}}}$ times more *blue shift* impact shall correspond to   an $f\left(\dfrac{1}{sec}\right)$; $f = \varphi(n)$ vibration value for unit period of time.

$c = 299,792$ km/sec is the quantum speed on the surface of the *Earth*.

Keeping *walkers* up flying is equivalent to the impact of *gravitation* as well.

It can be written therefore the following way:

6S1 $$\frac{dmc^2}{dt_i\varepsilon_x\sqrt{1-\dfrac{v^2}{c^2}}}\left(1-\sqrt{1-\frac{(c-i)^2}{c^2}}\right) - \frac{dmc^2}{dt_i\varepsilon_x}\left(1-\sqrt{1-\frac{(c-i)^2}{c^2}}\right) = \frac{dmc^2}{dt_i\varepsilon_E}\left(1-\sqrt{1-\frac{(g\Delta t)^2}{c^2}}\right)$$

where the left hand side of the equation represents the quantum impact relation of the oil bath and the walkers; the right hand side corresponds to the impact of *gravitation* – walkers have been quasi kept up, accelerated by *gravitation,* similarly as they have been placed on a support.

ε_x the intensity coefficient of the *Hydrocarbons* and ε_E of the *Earth* can be both taken equal to 1, as the electron process of the *Carbon* element is the one only communicating.

6S2 $$\frac{1}{\sqrt{1-\dfrac{v^2}{c^2}}}\left(1-\sqrt{1-\frac{(c-i)^2}{c^2}}\right) - \left(1-\sqrt{1-\frac{(c-i)^2}{c^2}}\right) = \left(1-\sqrt{1-\frac{(g\Delta t)^2}{c^2}}\right)$$

As $i = \lim a\Delta t = c$ and $\lim(c-i)$ corresponds to $(g\Delta t)$ in our time system

6S3 $$\left(1-\sqrt{1-\frac{(c-i)^2}{c^2}}\right)\left[\frac{1}{\sqrt{1-\dfrac{v^2}{c^2}}}-1\right] = \left(1-\sqrt{1-\frac{(g\Delta t)^2}{c^2}}\right); \qquad \frac{1}{\sqrt{1-\dfrac{v^2}{c^2}}} = 2$$

6S4   and $\dfrac{1}{2} = \sqrt{1-\dfrac{v^2}{c^2}}$; and $\dfrac{1}{4} = 1-\dfrac{v^2}{c^2}$; and $\dfrac{3}{4} = \dfrac{v^2}{c^2}$; and $v = c\sqrt{\dfrac{3}{4}}$; $v=286,000$ km/sec

For having the internal electron process *blue shift* impact of the oil bath *doubled,* equivalent to the speed increase of the mix at *v=286,000 km/sec* the vibration impact shall be of frequency $f = \varphi(n)$, function of *n* and the volume of the bath.

Basic info on the bath:
The type of the elementary process of the bath shall be selected. It shall be from those elementary compositions with natural electron process *blue shift* surplus: *hydrocarbons* or *water*.

In the case of *hydrocarbons* and *water* the *blue shift* conflict is coming from the elementary process of the *Hydrogen*, as $\lim \varepsilon_e^H = \infty$.

While the neutron processes of the *Hydrogen* process have been driven by the electron process *blue shift* surplus of the *Oxygen* process (in the case of water) or of the *Carbon* process (in the case of *Hydrocarbons*), the electron process of the *Hydrogen* process is in fact still acting (as of infinite low intensity) – ready for communication – for conflict.

Water is with balanced *Oxygen* proton/neutron processes and with proton process and electron process *blue shift* surplus of the *Hydrogen* process. The balanced status of the *Oxygen* process is coming from the fact, that *Oxygen* is using its electron process *blue shift* surplus and proton process cover driving neutrons of the elementary structure of the *Hydrogen* process.

The process is similar in the case of *Hydrocarbons*; just the balance has been established by the proton/neutron processes of the *Carbon* process. The neutron processes of the *Hydrogen* process has been driven by the *Carbon* process *blue shift* surplus and covered by the proton process of the *Carbon* process.

Comparing *water* and *hydrocarbons*, the *proton-electron* process component of the *Hydrogen* process available for *blue shift* conflict is more in *hydrocarbons* than in *water*: The electron processes of the *Carbon* process drive more their own neutron processes than the *Oxygen* process does. (The reason is that the electron process *blue shift* surplus of the *Oxygen* process is more than the intensity of the *Carbon* process.)

Ref.
6T2
6T3

This way, more *Oxygen* electron process *blue shift* surplus drives and takes more *Hydrogen* neutrons of infinite low intensity. At the same time, with reference to 6T2 and 6T3 the intensity of the electron process of the *Oxygen* process is less than that of the *Carbon* process.

The more value electron process *blue shift* conflict within the *hydrocarbons* and the higher *IQ* of the *hydrocarbons* explains why *hydrocarbons* are lighter than water. This is also the reason why the acting elementary *Hydrogen* process with electron process *blue shift* to be impacted by vibration is more within the *mineral oil* than within the *water*. The resulting effect is higher.

$\dfrac{dmc_O^2}{dt_i \varepsilon_{Oxigen}}\left(1-\sqrt{1-\dfrac{(c_O-i_O)^2}{c_O^2}}\right)$; and	$\dfrac{dmc_C^2}{dt_i \varepsilon_{Carbon}}\left(1-\sqrt{1-\dfrac{(c_C-i_C)^2}{c_C^2}}\right)$	$c = 299{,}792$ km/sec for both above the *Earth* surface;
$\dfrac{c_O^2}{\varepsilon_O}=8.847\text{E}+10$;	$\dfrac{c_C^2}{\varepsilon_C}=8.881\text{E}+10$	$\varepsilon_{Oxygen}=1.01533$ $\varepsilon_{Carbon}=1.01338$

6T1

6T2
6T3

Data have been calculated for all elementary processes and taken

The elementary process of the *Hydrocarbon* with vibration impact is ready for the conflict with *Earth gravitation* as well.

The intensity differences of the *water* and the *Hydrocarbons* are well demonstrated by their relation with fire.
Fire on the *Earth* surface is electron process *blue shift* conflict of infinite high intensity at $c = 299{,}792$ km/sec, the quantum speed of the *Earth* surface.

The quantum communication of the *water* impacted by *fire* intensifies the electron process *blue shift* drives of the processes. The *Hydrogen* process, impacted by *fire*, while its neutron process is still remaining of infinite low intensity, initiates the contribution of the external *Oxygen* processes: takes off the electron process of the *Oxygen* process from the fire – by that extinguishing fire.

$$e_e = \frac{dmc^2}{dt_i \varepsilon_e \sqrt{1-\dfrac{v^2}{c^2}}}\left(1-\sqrt{1-\frac{(c-i)^2}{c^2}}\right); \quad \text{as} \quad \varepsilon_e = \frac{dt_n}{dt_p}=\frac{\varepsilon_p}{\varepsilon_n}\sqrt{1-\frac{(c-i)^2}{c^2}}$$

6T4
6T5

The impacted elementary process becomes of less *blue shift* surplus and without conflicting intensity.

The description in speeding up terms is:

6T6
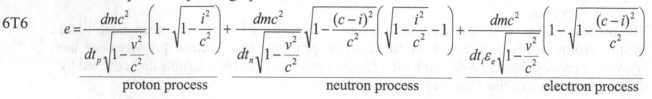

$$e = \frac{dmc^2}{dt_p\sqrt{1-\frac{v^2}{c^2}}}\left(1-\sqrt{1-\frac{i^2}{c^2}}\right) + \frac{dmc^2}{dt_n\sqrt{1-\frac{v^2}{c^2}}}\sqrt{1-\frac{(c-i)^2}{c^2}}\left(\sqrt{1-\frac{i^2}{c^2}}-1\right) + \frac{dmc^2}{dt_i\varepsilon_e\sqrt{1-\frac{v^2}{c^2}}}\left(1-\sqrt{1-\frac{(c-i)^2}{c^2}}\right)$$

$$\underbrace{\qquad\qquad}_{\text{proton process}} \qquad \underbrace{\qquad\qquad\qquad}_{\text{neutron process}} \qquad \underbrace{\qquad\qquad}_{\text{electron process}}$$

In other words, *fire* as external energy source is driving the water process – and for this cost extinguishing fire.
In sufficient volume this is enough for extinguishing the *fire*, otherwise as result of the developing conflict the water source just steaming away.

The temperature of the water and its consistency is well representing the balance.

The impact of *fire* on *Hydrocarbons* is similar, but the process is different. There is no external source around the fire for compensating the increasing intensity demand of the *Carbon* process.
The intensity increase of the electron processes of the *Carbon* and the *Hydrogen* processes even more deepens the conflict and in fact further feeds the *fire*.
The increasing intensity balance conflict also can create *water* drops in the fire as the *Hydrogen* process of *Hydrocarbons* may use the electron process and the proton cover of the *Oxygen* feeding the *fire*.

Rehabilitation of KOH by decontamination mix

The half-life of the *Potassium-40* (*Kalium-40*) isotope was reduced by the effect of a certain mix of especially selected elementary processes from <u>1.3 billion years to a couple of minutes</u>!

KOH was used for the experiment, as this is the only free for purchase chemical product.
There were 2 variants for studying the effect:

Experiment 1
    KOH in crystal format and the *decomix* in powder format are well mixed in glass pots.
    The activity of the mix is measured from above through the open cover of the glass pot:

1/1		1/2

GM measurement
 	
 ↑   ↑
 Mix: 1/1
 KOH +
     decomix
	After the mixing of the components *intensive heat generation* was detected in both cases. Heat generation is more intensive in case A/1. Activity remains reduced.
	 GM = Geiger-Muller device
		GM measurement
		
		 ⇑
		 Mix: 2/1
		 KOH +
		     decomix

Experiment 2:
    The plastic pot containing 1kg *KOH* crystals is placed into a larger vessel with *decomix*

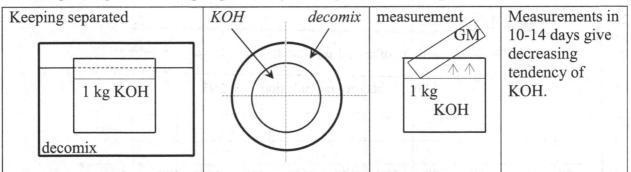

Keeping separated	*KOH* *decomix*	measurement GM	Measurements in 10-14 days give decreasing tendency of KOH.
1 kg KOH decomix		1 kg KOH	

Large number (80-130) of measurements have been taken for a single radiation data in both cases as the activity level is extremely low.

Activity of the KOH

Background radiation

Rehabilitation of KOH in the mix with the *decontamination mix*

Date of the measurement	Back-ground ρ_{bg} cpm	Portions: KOH–1 Dmix–1 ρ_{mix} cpm absolute values	Portions: KOH–1 Dmix–1 delta $\Delta\rho_{mix} = \rho_{mix} - \rho_{bg}$ Δ cpm	Portions: KOH–2 Dmix–1 ρ_{mix} cpm absolute values	Portions: KOH–2 Dmix–1 delta $\Delta\rho_{mix} = \rho_{mix} - \rho_{bg}$ Δ cpm	original KOH at the starting back-ground ρ_{KOH} cpm	days
22.10.2014 the date of mixing	15.18	18.37	immediate reduction to 3.19	20.83	immediate reduction to 5.66	40.21	0
05.11.2014	15.30	16.60	1.30	21.74	5.69		13
28.11.2014	15.97	18.47	2.50	22.94	6.96		37
05.01.2015	15.80	17.98	2.18	22.41	6.61		75
02.03.2015	16.03	17.52	1.49	22.32	6.29		132
21.04.2015	15.95	19.75	3.80	22.41	6.46		182
27.06.2015	16.36	20.21	3.85	20.21	6.12		243

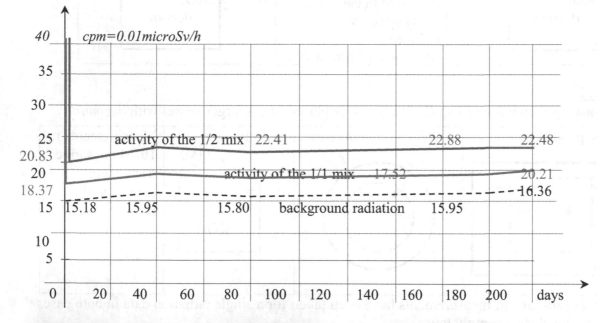

The *first* picture well illustrates the "home" conditions of the experiment.

The *second* and the *third* are especially selected from the huge number of data for the 1/1 and the 2/1 KOH/Dmix relations of the mix.

Rehabilitation of KOH in *decontamination* bath

1kg KOH is placed into a bath with 5kg Dmix Date	activities						
	Back-ground ρ_{bg} cpm	KOH in *Dmix* bath ρ_{KOH} cpm absolute values	KOH in *Dmix* bath delta $\Delta\rho_{KOH} =$ $= \rho_{KOH} -$ $- \rho_{bg}$ Δ cpm	KOH taken out of the bath ρ_{KOH} cpm absolute values	KOH taken out of the bath delta $\Delta\rho_{KOH} =$ $= \rho_{KOH} -$ $- \rho_{bg}$ Δ cpm	clean KOH ρ_{KOH} cpm absolute value	days
04.09.2014	16.60	**42.00**					0
10.09.2014	16.60	37.56	20.96	40.40	23.80	**42.00**	6
06.10.2014	16.60	34.55	17.95	35.28	18.68		32
06.11.2014	15.30	34.64	19.34	32.55	17.35		63
05.01.2015	15.80			32.52	16.72		123
02.03.2015	16.03			31.50	15.47		180
21.04.2015	15.95			32.64	16.69		230
27.06.2015	16.36			32.85	16.49		291

The first picture is the pot with KOH in the *decomix* bath.
The second and the third are one of the measurements within and out of the *decomix* bath.
The values are especially selected for the diagram as those are in vide range.

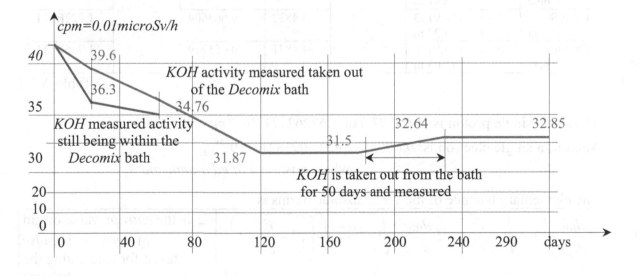

$cpm=0.01microSv/h$

KOH activity measured taken out of the *Decomix* bath

KOH measured activity still being within the *Decomix* bath

KOH is taken out from the bath for 50 days and measured

**Annex
7.2**

Isotope Rehabilitation by *Decontamination mix* – on the example of *Caesium-137*

Caesium has plenty of isotopes.
The followings are the ones around the stable status of *Cs-133*.

	PN	Atomic Weight/ Half-life	Proton/ Electron weights	Neutron weight	ε_x Intensity coefficient	c_x quantum speed	$IQ = \dfrac{c_x^2}{\varepsilon_x}$
Cs-129 +beta	55	128.9060 /32.06 h		73.4759	0.753986		1.691E+11
Cs-130 +beta	55	129.9067 /3.46 m		74.4766	0.743856		1.714E+11
Cs-131 +beta	55	130.9054 /9.689 d		75.4753	0.734012		1.738E+11
Cs-132 +beta	55	131.9064 /6.48 d		76.4763	0.724405		1.761E+11
Cs-133 =	**55**	**132.9054** /stable	**55.39985** /0.03025	**77.4753**	**0.715064**	**357,139**	**1.784E+11**
Cs-134 −beta	55	133.9067 /2.065 y		78.4766	0.705941		1.807E+11
Cs-135 −beta	55	134.9059 /2.3mill. y		79.4758	0.697065		1.830E+11
Cs-136 −beta	55	135.9073 /13.16 d		80.4772	0.688392		1.853E+11
Cs-137 −beta	55	136.9070 /30.167 y		81.4769	0.679945		1.876E+11
Cs-138 −beta	55	137.9110 /2.91 m		82.4809	0.671669		1.899E+11
Cs-139 −beta	55	138.9133 /9.27 m		83.4832	0.663604		1.922E+11
Cs-140 −beta	55	139.9172 /1.03 m		84.4871	0.655719		1.945E+11

Table
A.7.1   Table A.7.1.

Mass of a single proton is: *1.00727 u* or $1.67262171 \cdot 10^{-27}\,kg$

Mass of a single electron is: *0.00055 u* or $9.10938215 \cdot 10^{-31}\,kg$

where *u* is the *unified atomic mass*

The elementary balance of the *Cs* in absolute terms is:

7H1

$$\frac{dmc^2}{dt_p \varepsilon_p}\left(1 - \sqrt{1 - \frac{i^2}{c^2}}\right) = \xi \frac{dmc^2}{dt_n \varepsilon_n}\sqrt{1 - \frac{(c-i)^2}{c^2}}\left(\sqrt{1 - \frac{i^2}{c^2}} - 1\right);$$
$$\xi = 1$$

ξ is the *entropy* value origin of *quantum impulse* taken for calculating the balance

Elementary balance in intensity terms is:

7H2

$$\frac{dmc^2}{dt_p}\left(1 - \sqrt{1 - \frac{i^2}{c^2}}\right) \neq \frac{dmc^2}{dt_n}\sqrt{1 - \frac{(c-i)^2}{c^2}}\left(\sqrt{1 - \frac{i^2}{c^2}} - 1\right)$$

The quantum drive of the neutron collapse is the electron process:

$$\frac{dmc_x^2}{dt_i \varepsilon_x}\left(1 - \sqrt{1 - \frac{(c_x - i_x)^2}{c_x^2}}\right)$$

7H3

c_x the quantum speed and ε_x the intensity coefficient of the electron process - specifics of the elementary process. These two parameters establish the element.

There is an elementary process, with reference to Table 1.1 of Section 1, with intensity value of $\varepsilon_x = 0.688$ (as for the Cs-136 isotope), but

Ref.
Table
1.1

with speed of quantum communication equal to $c_x = 364,107$ km/sec.

This is *Ytterbium*.

The example from the other end is *Indium*, with $\varepsilon_x = 0.754$ (similarly to Cs-129), but

with the speed of quantum communication $c_x = 347,732$ km/sec.

This is for demonstrating, the *IQ* values of the *Caesium* process within Table 1 of the current annex.

Ref.
Table
A.7.1
Ref.
7H1

The deviation of the elementary balance of the *Caesium* isotope, with reference to 7H1 means:

$$\frac{dmc^2}{dt_p \varepsilon_p}\left(1 - \sqrt{1 - \frac{i^2}{c^2}}\right) \neq \xi \frac{dmc^2}{dt_n \varepsilon_n}\sqrt{1 - \frac{(c-i)^2}{c^2}}\left(\sqrt{1 - \frac{i^2}{c^2}} - 1\right)$$

7I1

The *IQ* of the *Caesium* process is damaged.

For the *Caesium* isotopes with damaged balance
- with less atomic weight than the naturally balanced *Cs-133*
  the intensity of the available electron process *blue shift* drive is more than needed.
  ($IQ_{Cs-133} > IQ_{Cs-129}$: as 1.784E+11> 1.691E+11).

    This is the reason these isotopes release electron process *blue shift* impact ($+\beta$)!
- with more atomic weight (like Cs-134, 135, 136, 137...)
  the available electron process *blue shift* drive is less than it is needed by the damaged elementary structure of the isotope.
  ($IQ_{Cs-137} > IQ_{Cs-133}$: as 1.876E+11>1.784E+11).

    This is the reason these isotopes are in electron process *blue shift* need ($-\beta$)!

Caesium isotopes as elementary processes try to re-establish the elementary balance therefore they are releasing electron process *blue shift* impact ($+\beta$) and proton process cover to, and are taking electron process *blue shift* impacts ($-\beta$) and proton process cover from the environment.

> Elementary evolution produces isotopic elementary structures and is in constant re-establishment of the global balance.

Decontamination by a *Decontamination mix* means, the mix of specific elementary processes is providing the necessary electron process *blue shift* drive and proton process cover to the damaged isotope!

The *Decontamination mix is* a specific composition of minerals with integrated electron process *blue shift* surplus and proton process dominance. The minerals of the elements with electron process *blue shift* surplus are: *O, H, N, C, S, Ca, Si*. The other components are from minerals with *Mg, Al, K, Na, Cl, Ti, Fe* elementary processes with increased intensity of the electron process *blue shift* impact and with dominant neutron process.

SA
7.2.1.
Ref.
7H1
7H2

A.7.2.1. <u>Rehabilitation</u>

With reference to the damaged balance in 7H1 of the *Cs-137*:

$$\frac{dmc^2_{Cs}}{dt_p\varepsilon_p}\left(1-\sqrt{1-\frac{i^2}{c^2}}\right) \neq \xi\frac{dmc^2_{Cs}}{dt_n\varepsilon_n}\sqrt{1-\frac{(c-i)^2}{c^2}}\left(\sqrt{1-\frac{i^2}{c^2}}-1\right)$$

Ref.
Table
1.

The absolute balance of the intensities of the proton and the neutron processes is damaged, as for the proton process it corresponds to its normal value, but for the neutron process (with reference to Table 1) it is increased.

The quantum speed of the *Caesium* process is: $c_x = 357,139$ km/sec;

Ref.
7H3

the intensity coefficient of the isotope (as per Table 1) and with reference to 7H3 below is of less value: $\varepsilon_{Cs-137} < \varepsilon_{Cs-133}$;

meaning: the Cs_{137} isotope needs an electron process *blue shift* drive with higher intensity impact!

7H3

$$\varepsilon_e = \frac{\varepsilon_p}{\varepsilon_n}\sqrt{1-\frac{(c-i)^2}{c^2}}$$ ε_p denotes the intensity of the proton,

ε_n the intensity of the neutron processes.

This way, the available electron process *blue shift* drive, corresponding to the balanced Cs_{133} is less in its capacity than it would be necessary for this specific *P/N* relation in Cs_{137}

$$e_{Cs-137} > e_{Cs-133}$$ – This is the reason of the isotope.

7H4

$$e_{Cs-137} = \frac{dmc^2_{Cs}}{dt_i\varepsilon_{Cs-137}}\left(1-\sqrt{1-\frac{(c_{Cs}-i_{Cs})^2}{c^2_{Cs}}}\right); \quad e_{Cs-133} = \frac{dmc^2_{Cs}}{dt_i\varepsilon_{Cs-133}}\left(1-\sqrt{1-\frac{(c_{Cs}-i_{Cs})^2}{c^2_{Cs}}}\right);$$

The deviation of the absolute balance of the proton/neutron process relation is:

7H5

$$\Delta_{137} = N_{137} - P_{137}$$

7I1

$$P_{137} = \frac{dmc^2_{Cs}}{dt_p\varepsilon_p}\left(1-\sqrt{1-\frac{i^2}{c^2}}\right); \quad \text{and} \quad N = \frac{dmc^2_{Cs}}{dt_n\varepsilon_n}\sqrt{1-\frac{(c-i)^2}{c^2}}\left(\sqrt{1-\frac{i^2}{c^2}}-1\right);$$

7I2

The *Decontamination mix* in absolute terms is in elementary balance state: $P_{Deco} = \xi \cdot N_{Deco}$ ($\xi = 1$ taken)

7J1

$$\frac{dmc^2_{Deco}}{dt_p\varepsilon_p}\left(1-\sqrt{1-\frac{i^2}{c^2}}\right) = \xi\frac{dmc^2_{Deco}}{dt_n\varepsilon_n}\sqrt{1-\frac{(c-i)^2}{c^2}}\left(\sqrt{1-\frac{i^2}{c^2}}-1\right)$$

while acting with electron process dominance, the description in intensity terms is:

7J2

$$\frac{dmc^2_{Deco}}{dt_p}\left(1-\sqrt{1-\frac{i^2}{c^2}}\right) > \frac{dmc^2_{Deco}}{dt_n}\sqrt{1-\frac{(c-i)^2}{c^2}}\left(\sqrt{1-\frac{i^2}{c^2}}-1\right);$$

Note: Elementary processes are acting in intensities.
The description in absolute terms is necessary for the demonstration of the balance.

7J3

$$e_{Deco} = \frac{dmc^2_{Deco}}{dt_i\varepsilon_{Deco}}\left(1-\sqrt{1-\frac{(c_x-i_x)^2}{c^2_x}}\right)$$

The available electron processes of the mix will drive the neutron processes of the *Cs-137* isotope – which is missing and needs electron process *blue shift* drive.

The *Decontamination mix* is composed from minerals with proton process dominance.

The electron processes of the *Decontamination mix* drive the neutron processes of the *Caesium-137* process and are covering them by proton process.

At the start of the rehabilitation: $P_{137} - N_{137} + P_{Deco} - N_{Deco} = -\Delta_{137}$ 7J4

At the end of the rehabilitation: $P_{137} - N_{137} + P_{Deco} - N_{Deco} = -\Delta_{Deco}$ 7J5

$$\text{as} \quad P_{Cs-133} = N_{Cs-133} \quad \text{and} \quad P_{Deco} - N_{Deco} = -\Delta_{Deco}$$ 7J6

The balance deviation of the *Cs-137* process is rehabilitating and it is turning into *Cs-133 process*.

SA 7.2.2

A.7.2.2. There is an important note to be taken into account

The electron process *blue shift* drive and the proton process cover of the *Decontamination mix* to the neutron process of the damaged *Cs-137* isotope is establishing that certain elementary process of the *Decontamination mix*, which is contributing to the rehabilitation of the isotope (one of the elementary processes introduced in Table 2.)

The reason is that neutron processes for certain extent are neutral.

The electron process *blue shift* drive and the proton process cover are establishing the *inflexion* of the formulating elementary processes.

This way, the neutron processes, having been damaged in the *Cs-137* isotope structure with insufficient electron process *blue shift* drive and proton process cover – once becoming driven and covered by the *Decontamination mix* – resulting in one of the elementary processes of the *Decontamination mix*.

The difference in the proton process intensity (*P)* and the neutron process intensity (*N)* of the isotope is decreasing.

The rehabilitation process however has its price:

the elementary processes of the *Decontamination mix* become damaged. With reference to 7J6, there will be a certain non-balance between their proton process and neutron process intensities, but this balance damage is rehabilitating in minutes or hours.

The half-life of the isotopes of *Decontamination mix:*

PN	Elements	isotope with the longest	half-life	The *decontamination mix* is capable to treat *alpha -*, and *gamma* type isotopes as well, as the principle of the rehabilitation is the same.
1	Hydrogen		level of 10^{-22} sec	
6	Carbon:	C-11	21 min	
7	Nitrogen:	N-13	10 min	
8	Oxygen:	O-15	2 min	
11	Sodium:	Na-25	1 min	
12	Magnesium:	Mg-28	21 hours	
13	Aluminium:	Al-29	7 min	
14	Silicon:	Si-31	2.5 hours	
16	Sulphur:	S-38	3 hours	
17	Chlorine:	Cl-38	38 min	
19	Potassium:	K-43	1 day	
20	Calcium:	Ca-45	163 days	
22	Titanium:	Ti-45	3 hours	
26	Iron:	Fe-59	45 days	

Table A.7.2

Table A.7.2

**Annex
7.3**

**Rehabilitation
of the *graphite* stacks, deforming and cracking in RBMK reactors**

SA.
7.3.1
    A.7.3.1. <u>About the reasons of the expansion of the *graphite* moderator, about its
deformation and the creaks of *graphite* stacks</u>

The *Carbon* elementary process is with proton process dominance.
The intensity of the proton process is more than the intensity of the neutron process.
    (The other elementary processes alongside with the *Carbon* of this certain life important group
    are the *Hydrogen, Helium, Oxygen, Nitrogen, Silicon, Calcium* and *Sulphur*.)
Carbon is specific elementary process of the nature – closest to the equal intensities of the
elementary proton/neutron process quantum communication, with the allotropes of
-    *diamond* representing the equal intensities itself,
-    *graphite* representing slight neutron process dominance with missing minor electron
    process *blue shift* impact, and
-    classical *carbon* is with the intensity dominance of the proton process, as the
    periodic table clearly proves it.

While *diamond* is with the quasi equally balanced intensities of the quantum
communication of the proton and neutron processes and having this way an extremely
stable elementary structure; *graphite* is with slight electron processes *blue shift* deficit and
neutron process dominance. This is the reason *graphite* process is less stable than *diamond*.
The internal structure of the *graphite* is soft.
The slight neutron process dominance would make *graphite* the highest grade of coal, as
the intensity of the electron processes of the *graphite* is higher than that is for the normal
Carbon process:

7K1
$$e = \frac{dmc^2}{dt_i \varepsilon_x}\left(1 - \sqrt{1 - \frac{(c-i)^2}{c^2}}\right) \text{ for } Carbon \text{ (coal) } \varepsilon_x > 1; \text{ for } graphite \ \varepsilon_x < 1$$

(The less is the value of ε_x, the intensity coefficient;
the higher is the intensity of the electron process.)
The advantage however at the same time is the disadvantage as well:
    graphite process is with intensive electron process *blue shift* impact but without
    surplus, therefore it is in fact impossible to ignite.

The *moderator* function of the *graphite* in RBMK reactors means generating electron
process *blue shift* conflict with the elementary process of the nuclear fuel in order to
initiate the fission. *Graphite* however will not be initiating *blue shift* conflict with the fuel
until it has not been in conflict itself.
Graphite is in quantum communication with the cooling *water*, which by its nature is with
electron process *blue shift* surplus and also in quantum communication with the fuel, which
either enriched by *U-235* or *Pu-239* are with increased proton process intensity and
electron process *blue shift* "surplus" relative to the normal elementary structure of the *U-
238* and the *Pu-244* processes – otherwise in their balanced state they are in deep
energy/mass intensity surplus of the neutron process.

The electron process *blue shift* surplus of the water and the intensity increase of the electron process of the fuel (even at this low level) are impacting the *graphite* process generating by this its internal *blue shift* conflict. The reason is the high sensitivity of the *graphite* process, as being so close to equilibrium.

As *graphite* is one of the allotropes of the *Carbon* process and as such is in quasi stable elementary process status, has no need in extra electron process *blue shift*. The experienced impact from the sides of the fuel and the water generates *blue shift* conflict within its internal quantum communication. The increased temperature is one of the symptoms of the generated electron process *blue shift* conflict (as consequence).

The heat generation in *graphite* as massive conflicting impact increases the already existing electron process *blue shift* impact intensity surplus within the fuel element, originally initiated only by water. The elementary structure of the fuel element cannot manage this increased electron process *blue shift* conflict and its internal balance becomes lost: it falls apart to elementary processes with less intensity characteristics.
The fission generates additional electron process *blue shift* conflict (heat), as the fission products, the generating new elementary processes are all with less intensities of the electron process than that for the enriched fuel element was. The cooling water takes the generating heat surplus away.
The establishing new elementary processes, results of fission however are all with damaged intensity balance of the proton/neutron processes – fission-product-isotopes.

The reason is that while the intensity of the electron process of the fuel is significantly high, the newly establishing elementary processes are with electron process *blue shift* demand –as the balanced elementary transition is missing.

$$e = \frac{dmc^2}{dt_i \varepsilon_{fuel}}\left(1 - \sqrt{1 - \frac{(c-i)^2}{c^2}}\right); \quad e = n\frac{dmc^2}{dt_i \varepsilon_{isotope}}\left(1 - \sqrt{1 - \frac{(c-i)^2}{c^2}}\right); \quad \varepsilon_{fuel} = \frac{\varepsilon_{isotope}}{n}; \qquad \text{7K2}$$

The less the value of the intensity coefficient of the electron process (ε_{fuel}, $\varepsilon_{isotope}$)

is, the higher is the intensity of the electron process.
The absolute balance of the proton and neutron processes is:

$$\frac{dmc^2}{dt_p \varepsilon_p}\left(1 - \sqrt{1 - \frac{i^2}{c^2}}\right) = \frac{dmc^2}{dt_n \varepsilon_n}\sqrt{1 - \frac{(c-i)^2}{c^2}}\left(1 - \sqrt{1 - \frac{i^2}{c^2}}\right); \qquad \text{7K3}$$

The value of the intensity coefficient is: $\quad \varepsilon_e = \frac{dt_n}{dt_p} = \frac{\varepsilon_p}{\varepsilon_n}\sqrt{1 - \frac{(c-i)^2}{c^2}} \qquad$ 7K4

Once for the isotopes of the fission product: $\varepsilon_{isotope} > \varepsilon_{fuel}$ 7K5

and as the summarised energy intensity of the proton processes of the newly establishing elementary processes is equal to the original intensity value of the fuel element, electron process *blue shift* drive is missing:

$$\sum \varepsilon_{p-isotope} = \varepsilon_{p-fuel} \qquad \text{7K6}$$

The consequence is missing neutron process intensities within the newly generating fission products – reason of the isotope status!

Neutron processes are *passive* processes. Neutron process needs electron process *blue shift* drive and should be covered by proton process.

The internal elementary balance problems of the generating fission-product-isotopes heavily impact the *graphite*, which is very sensitive (as earlier presented) to external quantum impact:

Quantum communication is ongoing, fission-product-isotopes are looking for electron process *blue shift* drive and proton process for covering the missing need.

Graphite process has been part of the quantum communication and as consequence is losing on the intensity of its electron process *blue shift* impact and proton process cover.

Graphite this way is providing two functions in RBMK reactors:

- as moderator is assisting to the fission, resulting in energy generation,
- **but** the fission of the *fuel* is also impacting it: fission-product-isotopes are taking electron process *blue shift* impact and proton process cover from the elementary structure of the *graphite* in order for covering their internal balance damage.

Losing on the elementary balance means the intensities of the proton and the electron processes are decreasing and as consequence the intensity of the (driven) neutron process of the elementary process of the *graphite* is decreasing the same way.

7L1
$$\frac{dmc_g^2}{dt_i\varepsilon_g}\left(1-\sqrt{1-\frac{(c_g-i_g)^2}{c_g^2}}\right) > \frac{dmc_{g-}^2}{dt_i\varepsilon_{g-}}\left(1-\sqrt{1-\frac{(c_{g-}-i_{g-})^2}{c_{g-}^2}}\right); \qquad \frac{c_g^2}{\varepsilon_g} > \frac{c_{g-}^2}{\varepsilon_{g-}}$$

Losing on the intensity of the electron process means: the value of the intensity coefficient of the electron process is increasing: $\varepsilon_{g-} > \varepsilon_g$

The lost quantum balance and the less intensity of the electron process drive result in the decrease of the intensities of the driven neutron and the anti-neutron processes as well.

As direct consequence (and in line with the controlling function), the intensity of the <u>anti-electron process becomes of increased value.</u> The reason is simple:

The relation of the intensities of the electron and anti-electron processes are reciprocal. If the intensity of the direct process is decreasing, the intensity of the anti-process should be more.

7L2
The other, "controlling-the-balance" explanation is that the intensity of the anti-electron process needs to be of increased value, as the intensity of the anti-proton process, driven by the anti-electron process shall be corresponding to the intensity of the proton process.

There are two points here to be noted:

(a) the meaning of the anti-processes

Processes and the anti-processes having common *inflexions* (like proton/anti-proton and neutron/anti-neutron) are always of equal intensities. Processes and anti-processes without *inflexion* (like electron/anti-electron processes) are in controlling function of events. If any deviation, the control process is initiating the correction.

7L3
(b) the specifics of the proton process.

The change of the intensity relation of the proton/neutron process – while the intensity of the proton process remains unchanged – means birth of isotopes.

The increase of the anti-electron process is a need, since the intensity of the proton process of the *graphite* is of standard value, characteristic of the elementary process that should be

Ref.
7L2
7L3
kept. The increased intensity of the anti-electron process ensures the collapse of the anti-proton process at the intensity level of the proton process expansion (with reference to the condition in 7L2 and 7L3). [The reason is the intensity loss of the electron process, the

drive of the neutron process – consequence of the quantum impact of the fission-product-isotopes.]

The increased intensity of the anti-electron process *blue shift* impact as above is natural need for assuring the intensity standard of the proton process:

$\varepsilon_{a-e} = \dfrac{\varepsilon_{a-n}}{\varepsilon_{a-p}} \sqrt{1 - \dfrac{(c-i)^2}{c^2}}$;	$\varepsilon_p = \varepsilon_{a-p}$; and $\varepsilon_{a-n} < \varepsilon_n$	The intensity of the anti-electron process is of increased value, as $\varepsilon_{a-e} < \varepsilon_e$	7M1

In the case of proton, neutron, anti-proton and anti-neutron processes, the value of the intensity coefficient is directly proportional to the real intensity of the process; the dt time component is the one reversely proportional to the intensity.

In the case of electron processes the time component is constant value: $dt_i = \dfrac{dt_o}{\sqrt{1 - \dfrac{i^2}{c^2}}}$; and $i = \lim a\Delta t = c$ 7M2

Ref 7M1

With reference to 7M1 above, the intensity increase of the anti-electron process *blue shift* impact means the increase of the intensity of the inter-elementary *Quantum Membrane* of the *graphite* process by the anti-direction, the controlling process. The taken by the fission-product-isotopes electron process *blue shift* impact weakens the intensity of the internal *Quantum Membrane* of the *graphite*. As direct consequence, the increasing intensity of the anti-electron process of the *graphite* tries to compensate it. The isotope demand however is constant; and the elementary balance of the *graphite* process is in permanent loss.

1. Alongside the appearance of the isotopes of the fission of the fuel, there are also appearing other types of isotopes generating now within the *graphite* itself, as the quantum impact of the fission is acting within the communicating *graphite* as well. All isotopes formulating within the *graphite* elementary structure are with electron process *blue shift* demand and with anti-electron processes of increased intensity.

2. *Graphite* originally is of slight neutron process dominance and also with slight anti-electron process surplus.

The reaction of the *graphite* is similar: quantum communication for assisting to re-establishing the balance. This deepens the conflict within the *graphite* stacks as the integrated intensity difference within the *Quantum Membranes* is further increasing: From one side (from the controlling anti-electron process) the intensity is increasing, from the other is taken more (fission-product-isotopes).

As consequence the internal *Quantum Membrane* of the elementary process of the *graphite* is expanding, leads to micro integrity problems, deforming and cracking the *graphite* stacks.

It might happen that *beta(-)* and *gamma* radiations as symptoms are escorting the appearance of the isotopes and the increasing intensity of the anti-electron process *blue shift* impact within the *graphite* moderator.

The appearance of isotopes within the *graphite* is the signal that the electron process *blue shift* intensity as of moderator function is decreasing. Massive appearance of isotopes within the *graphite* stacks – obvious results of operation – would also increase the *beta(-)* and *gamma* radiation impacts to the environment, as the increased than normal anti-electron process *blue shift* impact is affecting the *Quantum Membrane* and might generate electro-magnetic impact.

SA
7.3.2
Para.
a/

A.7.3.2. <u>Rehabilitation</u> of the <u>graphite</u> stacks

a./ <u>Off-line rehabilitation</u> of graphite stacks is relatively easy

After the mechanical repair and the correction of all mechanical damages, *graphite* stacks shall be positioned between columns of *Silicon-Calcium-Sulphur* mineral from all sides. This is necessary as the mechanical treatment will not free the *graphite* stacks from the accumulating isotopes within.

Silicon and the other elementary process are with proton process dominance and natural electron process *blue shift* surplus – similar to the *Carbon* process.
These columns surrounding *graphite* stacks communicate at quantum level with the damaged elementary processes within the *graphite* structure:
The minerals provide the missing electron process *blue shift* drive and proton process cover to the neutron processes of the isotopes.
By providing the missing quantum impact, the elementary process of the minerals will be damaged, but the half-life of the self-rehabilitation of the processes is dimension of hours. (*Calcium* is 60 days.)
The process of *graphite* rehabilitation between *mineral* columns can be followed and controlled by *radiation* measurement.
This is the case when radiation data as "benefit" could be used.
As result of the rehabilitation, *graphite* will be with certain content of the "healing" minerals. (The rehabilitation impact from the minerals results in the elementary process of the mineral.)
The *thermal conductivity* of *Silicon* for example however is around 149 $W \cdot m^{-1} \cdot K^{-1}$.
From practical point of view this is equal/similar to the *thermal conductivity* of the *graphite* stacks, which is about 119-165 $W \cdot m^{-1} \cdot K^{-1}$. This similarity proves, there is no thermal risk for fulfilling the moderator function in RBMK reactors.

Para.
b./

b./ <u>Possibility of <u>on-line rehabilitation</u> and <u>protection</u></u>

In the case of successful experimental proof of the off-line rehabilitation of *graphite* stacks by the elementary process of *Silicon* and/or other minerals,
detailed thermal/nuclear assessments, calculations can be made on the possibility of the online protection of *graphite* stacks during operation – without dismantling of the RBMK reactors!
Mineral rods could be positioned between the *graphite* stacks of the RBMK active zone, resulting also in the elementary process of the minerals within the RBMK construction.
The selected minerals also would be expected working as moderator, as their elementary processes are with electron process *blue shift* surplus and proton process cover. The minerals would contribute to keep the *blue shift* conflict up, without reducing the fission potential of fuel.
The on-line direct quantum communication of the elementary processes of the minerals with the *graphite* moderator and with the formulating fission-product-isotopes reduces the damaging load of the *Quantum Membrane* of the *graphite*.
There is a good chance that *graphite* stacks expansion, deformations and cracks will be stopped and the reactor can operate safely without *graphite* moderator damages.
The selected minerals provide the necessary electron process *blue shift* impact and proton process cover to the isotopes of the fission and to the *graphite* process.

The *pyramid* replica experiment
The results of the measurements with the *pyramid* replica

The objective of the measurements is to demonstrate the energy potential of the *pyramid*.
The *pyramid* replica – with base 785.4 mm and height 500 mm – is the true copy of the
Great Pyramid of Giza (with 230 m and 146.5 m). The principle of the generating energy
potential within the *Giza pyramid* and the replica is similar. Just the *Giza* and the other
pyramids do not have the wires inside and the cable outside for measuring it.
The mass relation between the *Great Pyramid* and the replica is 25.78 million/1!
The measurements were made in the continental climate conditions of Hungary in March
and April 2015.

Wires were built in the concreate structure
with a measurement terminal connected to
the soil for measuring the developing
potential within the replica.

Two temperature indicators were built in
at distance 1/3 of the height from the top
and from the bottom
– for measuring the temperature increase of
the internal conflict.
The temperatures of the surface at these
heights were also measured

The temperature of the air around the
replica was measured.

Temperature measurements
(1) on the surface of the soil,
(2) at 200 mm depth around the *pyramid*
and (3) 200 mm below the *pyramid*.

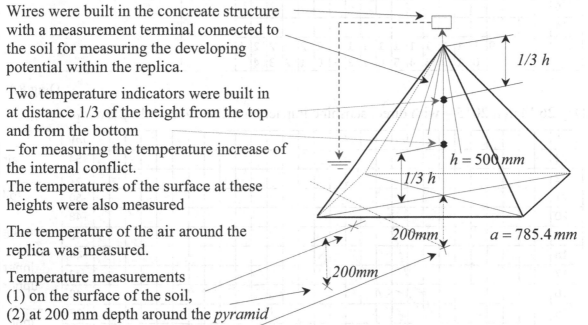

The building material of the replica is a certain composition of minerals with elementary
processes of *O, H, N, C, S, Ca, Si, Mg, Al, K, Na, Cl, Ti, Fe*. The unique characteristic of
this solid mineral composition is its electron process *blue shift* surplus.
With reference to Section 8, this elementary composition generates *blue shift* conflict
under the anti-electron process *blue shift* quantum impact of *gravitation*.
The measured data and the diagrams of the results are given in the following pages.
The pictures at the end of the annex were made during the experiment.

Ref.
S.8

Connecting the cable contact at the top of the pyramid replica with the *Earth* surface the
measured voltage is a value around *+200 mV*, sometimes reaching *+250 mV* and more. The
value depends on the pressure the *Earth* contact is positioned into the soil, depth of 2-5 cm.
With higher pressure the voltage value goes up and only measurable if the scale of the
device is changed. With less pressure and less deepness the voltage is around *180 mV*.

Day 25 March 2015 – without direct sunshine impact on the casing of the *pyramid*

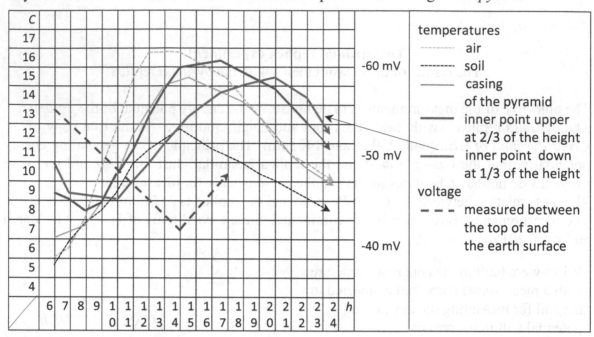

Diag. 8A1

Diag.8A1

Day 26 March 2015 – with direct sunshine impact on the surface of the *pyramid*

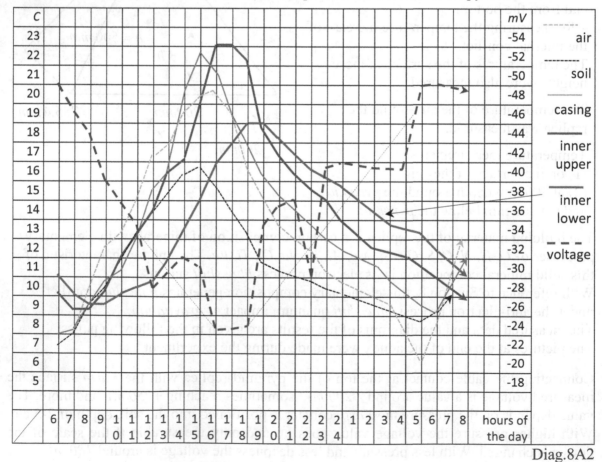

Diag. 8A2

Diag.8A2

day/ hours	Temperatures °C								Voltage between the top and the soil mV
	air	soil		below 400 mm	pyramid casing		pyramid inside		
		around			upper part at 1/3	lower part at 2/3	at 1/3 from the top	at 1/3 from the bottom	
		surface	in 200 mm						
25.03									
06:00	6.0	5.8	7.2	8.0	7.0	7.3	9.6	11	-55.4
07:00	6.5	6.8	7.1	8.0	7.4	7.4	8.1	9.5	-50.3
08:00	8.0	7.7	7.3	7.9	7.8	7.8	8.4	9.5	-47.2
09:00	10.7	8.3	7.1	8.0	8.9	8.9	8.8	9.1	-45.8
10:00	12.8	9.3	8.1	7.8	10.5	10.5	10.2	9.0	-42.7
11:00	15.1	11.1	7.6	8.8	12.2	12.2	11.3	9.9	-44.3
12:00	16.8	11.6	8.1	7.8	13.8	13.1	13.5	10.8	-43.9
13:00	16.8	12.1	8.2	7.8	15.0	14.6	14.9	11.9	-44.3
14:00	16.7	12.7	8.4	7.8	15.5	14.7	15.9	12.8	-44.7
14:30	16.5	12.1	8.6	7.8	15.6	14.8	16.1	13.3	-42.9
16:30	14.4	11.5	9.2	8.7	14.7	14.1	16.2	14.8	-48.6
17:00	14.3	11.1	9.3	8.6	14.4	13.9	15.8	14.7	-49.7
18:00	13.2	10.9	9.3	8.6	14.1	13.6	15.5	14.9	
19:00	12.9	10.3	9.3	8.6	13.2	13.2	15.5	14.8	
20:00	10.8	10.0	9.3	8.6	12.3	12.2	14.8	15.3	
21:00	10.5	9.6	9.6	8.6	11.3	11.2	13.0	13.8	
22:00	10.0	9.3	9.3	8.9	10.8	10.8	13.2	14.1	
23:00	9.4	8.8	8.8	8.7	10.2	10.2	12.1	13.0	
22:30	9.1	8.4	8.4	8.6	9.8	9.8	11.5	12.6	
26.03.									
06:00	7.5	7.5	7.9	8.4	8.1	8.2	10.1	11.1	-50.1
07:00	8.0	8.1	7.9	8.4	8.3	8.4	9.3	10.1	-45.8
08:00	10.3	9.5	8.1	8.4	9.6	9.6	9.3	9.6	-42.8
09:00	12.2	10.2	8.6	8.4	10.6	10.2	10.3	9.5	-37.6
10:00	13.1	10.5	8.2	8.4	11.3	11.0	11.0	9.9	-36.5
11:00	15.0	13.2	8.8	8.8	12.9	12.3	12.0	10.3	-33.6
12:00	17.3	14.5	9.1	8.8	15.4	14.4	13.2	10.4	-28.7
13:00	17.9	14.6	9.4	8.7	16.2	15.0	15.2	11.8	-30.7
14:00	19.4	16.2	10.3	8.5	20.8	19.2	16.2	12.0	-32.3
15:00	20.2	16.8	10.6	8.5	22.8	20.8	19.8	15.0	-31.2
16:00	20.7	15.5	11.5	8.6	21.5	19.0	22.1	16.9	-24.6
17:00	19.5	14.4	11.4	8.7	19.8	18.5	21.9	17.9	-23.8
18:00	16.8	13.1	11.3	8.7	18.0	16.9	21.1	18.9	-24.0
19:00	15.2	11.9	11.3	9.0	16.6	16.1	18.9	18.8	-34.8
20:00	13.7	11.6	11.1	9.2	15.5	15.4	17.5	18.2	-36.8
21:00	13.0	11.2	10.9	9.2	15.3	15.1	16.5	17.4	-37.6
22:00	12.3	10.9	10.6	9.2	14.3	14.3	15.5	16.5	-30.0
23:00	11.3	10.5	10.3	9.4	13.0	13.0	15.1	16.3	-41.0
00:00	10.9	10.2	10.3	9.4	13.0	13.0	15.1	16.3	-41.0
01:00									
02:00	11.4	10.1	10.1	9.8	12.5	12.5	13.9	15.4	-41.0
03:00	9.6	9.9	9.9	9.8	11.7	11.7	12.7	14.1	-42.1
04:00									
05:00	6.6	8.8	9.4	9.7	9.5	9.5	12.1	13.6	-47.6
06:00	8.4	9.2	9.3	9.7	9.2	9.2	10.6	12.1	-47.9
07:00	11.3	9.9	9.3	9.7	10.5	10.5	10.7	11.5	-46.3

Table 8A1

Day 1 and 2 April 2015 – with cold wind and sunshine

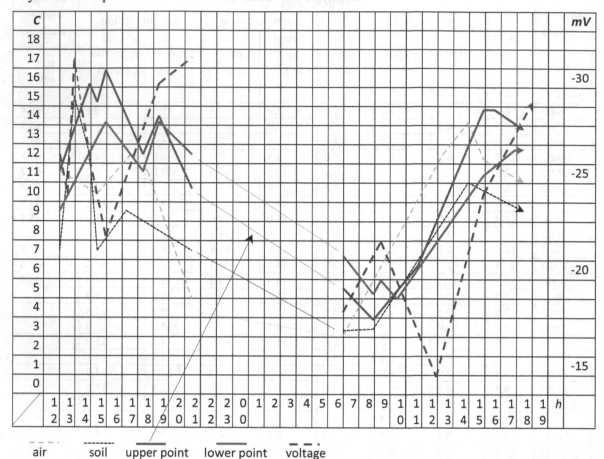

Diag.
8A3

air soil upper point lower point voltage

Diag.8A3

the summary of the measured data of Tables 8A2 and 8A3

day/ hours	Temperatures $^{\circ}C$						Voltage between the top and the soil *mV*	notes
	air	soil	casing		pyramid inside			
			up	down	upper	lower		
01.04								
12:00	11.4	7.6	15.2	9.9	**11.8**	**9.3**	**-26.6**	10 mm depth
12:30	11.3	7.3	15.3	9.7	**13.4**	**9.7**	**-24.5**	100 mm
13:00	11.0	15.7	15.3	9.9	**13.9**	**9.9**	**-30.6**	sunny place: surface
13:30	11.3	15.6	13.9	10.9	**15.1**	**10.9**	**-29.1**	sunny place: 10 mm
14:00	10.7	12.7	12.5	11.6	**16.1**	**11.6**	**-26.4**	sunny place: 100mm
14:30	10.3	7.6	12.6	12.2	**15.2**	**12.2**	**-21.4**	surface
15:00	10.9	8.6	14.2	14.0	**16.9**	**14.0**	**-22.1**	100 mm
16:30	12.4	9.5	15.2	12.9	**13.6**	**12.9**	**-25.7**	100 mm
17:30	11.3	9.3	13.5	11.8	**12.6**	**11.8**	**-28.6**	100 mm
18:30	8.7	8.7	10.2	14.0	**14.3**	**14.0**	**-30.1**	100 mm
19:30	7.4	8.2	9.9	13.5	**12.7**	**13.5**	**-30.8**	100 mm
20:30	5.0	7.6	7.7	12.7	**10.8**	**12.7**	**-31.4**	100 mm

The yellow means the temperature detector was re-placed to the place with direct sunshine impact.

Table
8A2

Table 8A2

day/ hours	Temperatures °C				Voltage between the top and the soil mV
	air	soil	pyramid inside		
			upper	lower	
02.04					
6:00	2.9	3.2	5.7	7.1	-18.1
7:30	4.0	3.2	3.8	5.2	-18.5
8:00	4.6	3.4	3.9	5.1	-20.6
8:30	6.0	4.1	4.7	5.5	-20.8
9:30	7.9	4.7	4.7	4.9	-18.1
10:45	10.6	5.9	5.9	5.2	-18.1
12:00	11.8	8.6	9.8	7.2	-14.9
12:30	12.4	8.9	10.8	7.6	-17.4
13:00	12.8	9.3	11.8	8.1	-19.2
14:00	14.1	10.9	13.1	9.2	-20.9
15:00	12.2	10.6	14.8	11.5	-24.4
15:30	12.2	10.5	14.8	12.2	-25.5
17:00	11.9	9.8	14.0	12.7	-27.4
17:30	11.0	9.6	13.8	12.7	-28.0

Important note relating all measurements:

Placing, positioning and moving the minus contact into or alongside or within the Earth surface results in plus voltage, value of **180-250 mV** each time.
This plus value voltage is changing with a gradient of 0.1-1 mV/sec – to the inflexion and after the minus voltage is building up and stabilising as the values within the table are given.
Plus voltage means electricity supply from the pyramid; minus voltage means gravitation feeds the pyramid through the top.

Table 8A3   Table 8A3

3 April – with extremely cold air temperature for this time of the year.

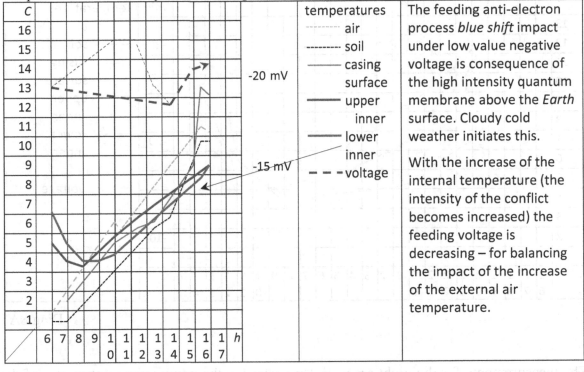

temperatures
-------- air
--------- soil
——— casing surface
——— upper inner
——— lower inner
– – – voltage

-20 mV

-15 mV

The feeding anti-electron process *blue shift* impact under low value negative voltage is consequence of the high intensity quantum membrane above the *Earth* surface. Cloudy cold weather initiates this.

With the increase of the internal temperature (the intensity of the conflict becomes increased) the feeding voltage is decreasing – for balancing the impact of the increase of the external air temperature.

Diag.8A4   Diag. 8A4

With reference to Diag.8A1-8A3 the feeding gravitation quantum impact is decreasing with the increase of the internal temperature.
The increasing internal temperature demonstrates the *gravitation* impact through the basic surface of the pyramid. The decreasing internal temperature initiates the increase of the value of the feeding voltage during night times.

day/ hours	Temperatures ^{o}C					Voltage between the top and the soil mV	Notes:
	air	soil	casing surface	pyramid inside			If after the measurement the connection between the top and *Earth* surface was interrupted, the accumulated potential release started at **180-250 mV**. The negative feeding voltage, marked in this table is building up from the inflexion with low value gradient.
				upper	lower		
03.04							
6:00	1.9	1.3		5.5	7.0	-19.0	
7:00	2.0	1.3	2.4	3.7	5.4	-20.1	
8:00	4.2	2.6	2.9	3.3	4.7	-21.1	
9:00	5.1	3.6	3.9	4.0	4.7	-21.8	
10:00	6.5	4.7	5.3	4.9	4.9	-22.2	In the case of permanent move, change of the Earth surface contact the positive voltage was vibrant and quasi stable at value around 200-**250 mV**.
10:30	6.0	5.3	5.6	5.9	5.5	-21.9	
11:30	6.9	6.3	6.3	6.4	6.4	-22.0	
12:30	7.4	6.4	7.5	6.9	6.9	-19.0	
13:30	8.2	6.6	8.8	7.6	7.6	-18.5	
15:00	8.8	9.8	9.3	7.9	7.9	-20.4	
15:30	11.3	10.8	13.5	8.9	8.9	-20.5	
16:00	11.0	10.7	12.8	9.3	9.3	-20.7	

Table
8A4

Table 8A4

Night of 6 to 7 April

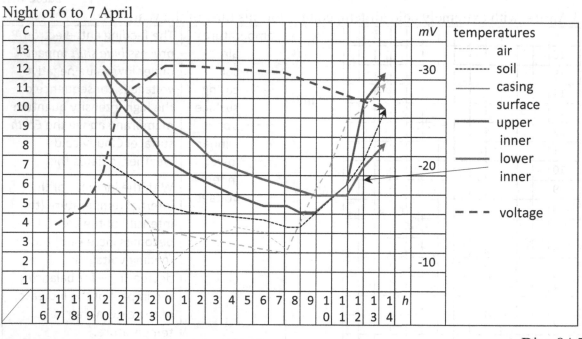

Diag.
8A5

Diag.8A5

The measurements for the night are well demonstrating the compensating character of the quantum impact of gravitation.

day/ hours	Temperatures °C					Voltage between the top and the soil *mV*
	air	soil	casing surface	pyramid inside		
				upper	lower	
06.04						
16:30						-14.5
17:30						-15.6
18:30						-16.5
19:30	6.5	7.8	8.3	12.3	12.6	-19.8
20:30	6.2	7.2	7.2	10.8	11.8	-26.5
21:30	5.3	6.8	7.2	10.0	11.2	-28.7
22:30	4.5	6.2	6.9	9.0	10.6	-30.3
23:30	2.1	5.3	5.2	7.8	9.8	-31.6
07.04						
1:00	3.2	5.0	4.8	7.0	9.0	-31.5
2:30	3.9	5.1	4.2	6.4	7.9	-31.0
3:45	4.3	4.9	4.5	6.0	7.4	-30.6
5:45	4.0	4.8	3.9	5.6	6.8	-30.7
7:15	3.2	4.2	3.6	5.5	6.7	-30.4
8:05	4.0	4.2	3.6	5.1	6.3	-30.1
9:00	6.1	5.0	4.7	5.1	5.9	-29.0
10:00	7.6	5.3	5.9	5.8	5.9	-28.4
11:00	9.9	6.3	7.8	6.7	6.0	-27.8
12:00	10.3	7.8	9.2	9.8	7.2	-29.0
13:10	11.9	10.4	12.0	12.1	8.6	-27.0

Comparing the measured temperatures at the higher and the lower positions of the internals, the higher is the one mainly reflecting the external air temperature impact; the lower is mainly about the quantum impact of gravitation to the quantum membrane of the pyramid.

The external temperature is extremely cold on late afternoon of the 6-th. Both temperatures are significantly higher than the air temperature.

The red line (of the lower position) is of higher value from late afternoon for all night than the blue one (the upper position), which takes over only in late morning of the 7-th.

Table 8A5

Table 8A4

Picture 8A1	Picture 8A2
Feeding from the *Earth* surface to the pyramid: -16.5 mV	Release of the voltage potential from the pyramid to the *Earth* surface: 257 mV

Pict.
8A1
8A2

Pic.
8A3
8A4

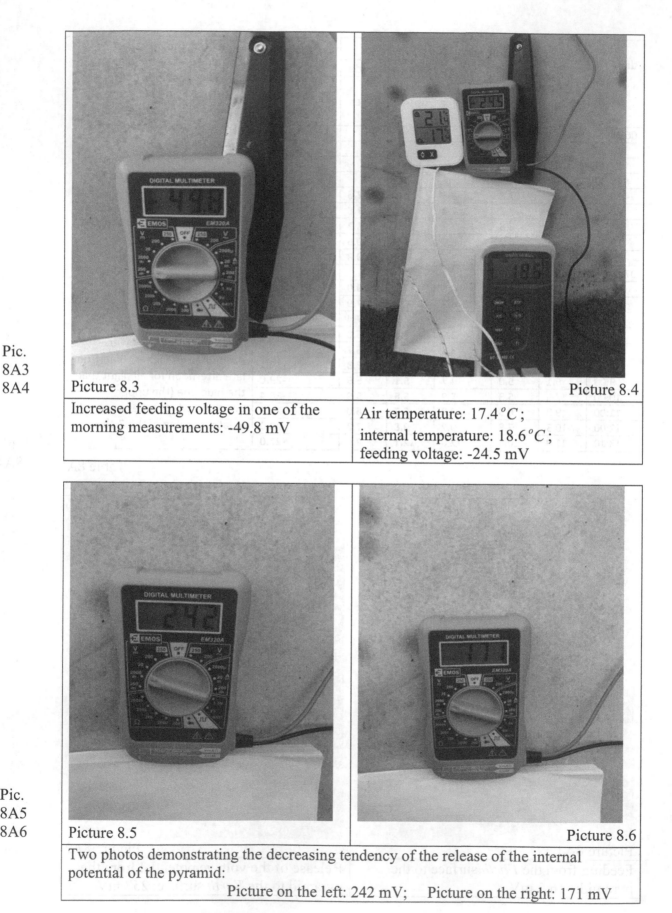

Picture 8.3

Increased feeding voltage in one of the morning measurements: -49.8 mV

Picture 8.4

Air temperature: $17.4\,^\circ C$;

internal temperature: $18.6\,^\circ C$;

feeding voltage: -24.5 mV

Pic.
8A5
8A6

Picture 8.5

Picture 8.6

Two photos demonstrating the decreasing tendency of the release of the internal potential of the pyramid:

Picture on the left: 242 mV;　　Picture on the right: 171 mV

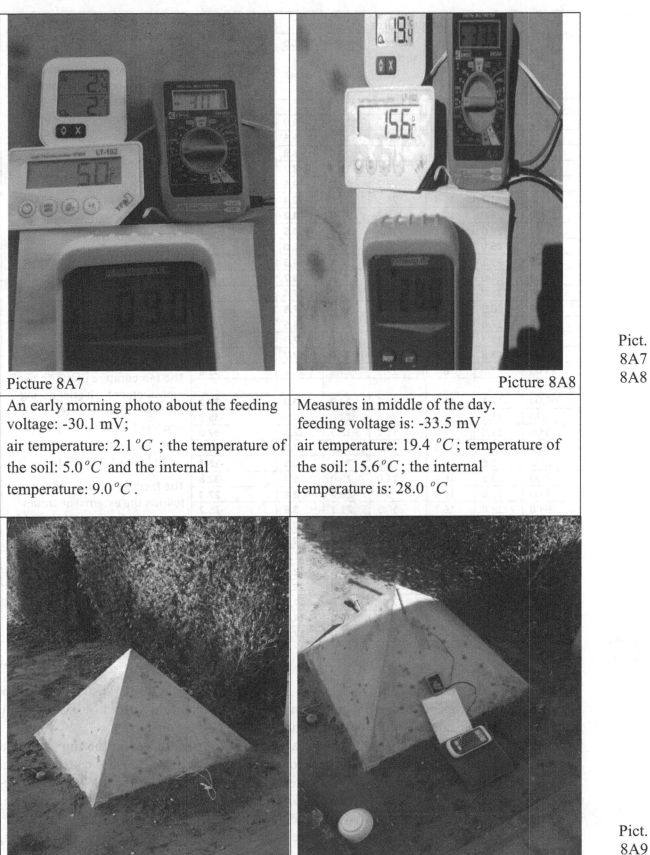

Picture 8A7	Picture 8A8
An early morning photo about the feeding voltage: -30.1 mV; air temperature: 2.1 $^\circ C$; the temperature of the soil: 5.0 $^\circ C$ and the internal temperature: 9.0 $^\circ C$.	Measures in middle of the day. feeding voltage is: -33.5 mV air temperature: 19.4 $^\circ C$; temperature of the soil: 15.6 $^\circ C$; the internal temperature is: 28.0 $^\circ C$

Picture 8A9, 8A10: The pyramid replica

Pict.
8A7
8A8

Pict.
8A9
8A10

day/ hours	Temperatures $^\circ C$					Voltage between the top and the soil mV	The temperature of the Earth below the pyramid in practical terms is without change. There is a slow increase/decrease depending on the part of the day, but it has no impact on the internal temperature of the replica.
	air	soil	casing surface	pyramid inside			
				upper	lower		
26.04							
7:45	14.1	15.8	15.0	21.1	9.9	+270.0	
8:10	15.0	15.7	15.2	20.3	9.3	+52.5	
9:00	17.4	15.6	16.0	20.0	8.8	+4.0	The temperature of the surface follows the air temperature.
10:00	18.6	15.6	17.0	23.2	8.7	-3.8	
11:00	20.7	15.6	23.7	27.3	9.5	-7.9	The internal temperatures are taking the quantum impact of the Sun, especially the "up" measurement. The up temperature for the whole period is higher than the external air temperature, which means the energy is taken as direct quantum impact from the Sunshine.
12:00	24.7	15.7	24.5	29.5	10.4	-10.5	
13:00	28.0	16.1	25.7	33.5	14.2	-9.4	
14:00	29.0	16.2	28.8	35.7	14.5	-3.7	
15:00	26.0	16.5	28.0	29.7	12.0	-5.6	
16:00	25.1	16.8	27.0	32.4	14.3	-4.6	
17:00	24.6	17.2	26.8	31.0	15.7	-3.8	
18:00	22.7	17.5	25.5	33.2	17.4	-4.6	
19:00	20.9	17.8	24.7	31.6	17.8	-7.6	
19:45	17.8	18.1	22.8	31.3	18.5	-12.4	
21:00	15.3	18.1	20.5	29.2	17.9	-20.0	The temperature of the "down" measurement is higher during the night than the external air and less for the whole period than the temperature of the Earth below the replica.
22:00	13.8	18.2	18.6	27.6	17.7	-26.2	
23:00	12.9	18.0	17.5	25.9	16.1	-31.8	
27.04							
00:00	12.4	18.0	16.8	24.9	15.0	-33.5	
1:20	11.5	17.8	15.5	23.8	14.3	-35.9	
3:00	11.8	17.5	15.2	22.3	12.3	-37.0	
4:30	11.7	17.1	15.0	21.4	11.3	-38.5	
6:00	10.6	16.8	14.0	20.7	10.3	-38.3	The feeding/loading voltage follows the external quantum impact. If the external impact is decreasing, the feeding impact from gravitation is compensating the loss. This is well demonstrated by the measurement in night time.
7:00	12.1	16.6	14.0	20.0	9.4	-32.8	
8:00	15.8	16.3	15.0	20.0	8.8	-27.9	
9:00	18.8	16.3	16.3	21.1	8.9	-25.2	
10:00	20.9	16.1	17.8	24.3	9.3	-22.6	
11:00	23.2	16.1	26.3	27.9	9.7	-19.9	
12:00	29.6	16.3	27.3	32.4	11.8	-18.9	
13:00	26.4	16.7	29.3	33.3	12.5	-18.9	
14:00	30.0	17.1	29.0	35.0	15.0	-12.3	
15:00	24.9	17.2	27.3	35.6	17.5	-11.4	
16:00	24.3	17.6	26.5	29.8	14.0	-9.4	
17:00	23.9	17.7	26.3	32.8	17.9	-8.9	

Table 8A6

Table 8A6

The diagram of the data of Table 8A6 on the next page clearly shows the distribution of the feeding quantum impacts of the *Sun* and *gravitation:* The quantum impact of gravitation is moving in the opposite direction to the quantum impact of the *Sunshine.*

Other diagrams also prove that while there are significant differences in the impact of the *Sunshine* as weather conditions might be cloudy or clear, and the conflict causing by the *Sunshine* impact vary depending on the external temperature, the quantum impact of the *Sun* is the main source during daily time. *Gravitation* is compensating and balancing the difference for having constant load within the pyramid.

In the case the intensity of the consumer side is increasing, the gravitation is compensating and covering the demand.

Day 26 and 27 April 2015 – with extremely high temperature

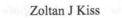

| t | 8 | 9 | 10 | 11 | 12 | 13 | 14 | 15 | 16 | 17 | 18 | 19 | 20 | 21 | 22 | 23 | 0 | 1 | 2 | 3 | 4 | 5 | 6 | 7 | 8 | 9 | 10 | 11 | 12 | 13 | 14 | 15 | 16 |
| h | | | | | | | | | | | | | | | | | | | | | | | | | | | | | | | | | U |

air ——— soil --------- upper point ━━━ lower point ━━━ surface ━━━ voltage ━ ━ ━

Diag.8A6

Diag.
8A6

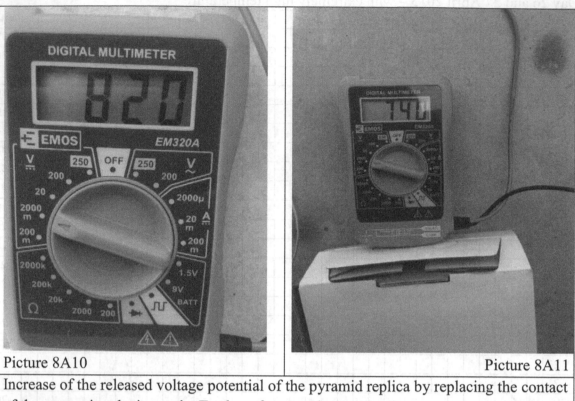

Pict.
8A10
8A11

Picture 8A10	Picture 8A11
Increase of the released voltage potential of the pyramid replica by replacing the contact of the measuring device to the Earth surface to *Aluminium* alloy.	
Picture on the right: 820 mV	Picture on the left: 740 mV

The objective of the experiment on 4. May was to take a video about the *inflexion* of the voltage released by the replica to the feeding by *gravitation*.
The measurement started at 10:15.

The external temperature was high, $25\,^{\circ}C$ in shadow. The casing has direct *Sunshine* impact.

Time 04/05 May	Temperature $^{\circ}C$			Voltage mV	
	air	inside upper	inside lower		The starting measured voltage was the usual high positive value. The decrease was extremely slow. Without registering the intermediate data, (as being always expecting the change) the voltage was even still late evening of positive value. The *inflexion* from release to feeding/loading by *gravitation* happened at around 22:00. Its maximum value was measured in the early morning next day. The feeding dominance lasted to a time point between 8:00 and 9:00.
10:15	25.0	36.0	18.0	250	
...				...	
20:30				2.4	
21:30	20.2	31.2	17.2	1.6	
22:30	17.9	29.6	16.4	-1.9	
23:00	17.9	28.8	15.9	-3.7	
23:30	17.7	28.4	15.6	-5.2	
5:00	16.2	25.9	12.4	-12.4	
6:00	16.0	24.5	11.7	-13.0	
*7:00	17.1	23.7	10.6	-12.6	
8:00	18.1	24.2	10.6	-2.1	
9:00	19.4	24.2	10.2	+0.9	
9:45	21.0	25.1	10.0	+2.6	

Table
8A7

* means the *Sun* shines through

Table 8A7

4-5 April measurements

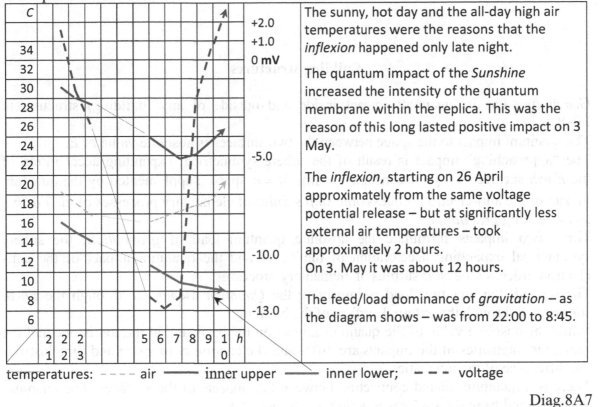

The sunny, hot day and the all-day high air temperatures were the reasons that the *inflexion* happened only late night.

The quantum impact of the *Sunshine* increased the intensity of the quantum membrane within the replica. This was the reason of this long lasted positive impact on 3 May.

The *inflexion*, starting on 26 April approximately from the same voltage potential release – but at significantly less external air temperatures – took approximately 2 hours.
On 3. May it was about 12 hours.

The feed/load dominance of *gravitation* – as the diagram shows – was from 22:00 to 8:45.

temperatures: - - - - air ——— inner upper ——— inner lower; – – – voltage

Diag.8A7   Diag. 8A7

The lesson is:
The mains source of the feeding is *gravitation,* but in sunny, hot days the direct and the indirect *Sunshine* impact increases the internal electron process *blue shift* conflict of the quantum membrane of the pyramid, initiated by the quantum impact of *gravitation*. The *Sunshine* quantum impact partially "substitutes" the quantum impact of *gravitation*.

The communication is quantum impact based, since the voltage potential of the pyramid is communicating with the *Earth* even through the turned off terminal of the measurement. If the contact is disrupted, the communication starts again from the highest potential of the pyramid.

**Annex
8.2**

Building structures

Gravitation has its quantum impact inside and outside of any (building) structure in Fig.8.3.

The quantum impact to the space between the two surfaces is absolute value of *E*.

The "approaching" impact is result of the sphere symmetrical expanding acceleration of the *Earth* surface, $a = g$ at constant $i = \lim a\Delta t = c$ speed; supplemented by the quantum impact of the anti-electron process *blue shift* surplus of elementary processes of the *Earth* – as for feeding *gravitation*.

These two impacts formulate the absolute quantum load of *gravitation*: the sphere symmetrical expanding acceleration of the *Earth* and the quantum impact of the anti-electron process *blue shift* surplus of elementary processes.

The second, the quantum load is impacting the *Quantum Membrane* through the *Earth* surface and through the external surface of the building.

While the absolute value of the quantum impact on the basis and at the top is one and the same, the intensities of the impacts are different. The continuity of space and time justifies the difference of the intensities.

There is a quantum related connection between the impacts of the surfaces. The quantum impact to and from the surfaces is acting in a balanced way.

This is the reason terminology "transfer" will be used in the followings.

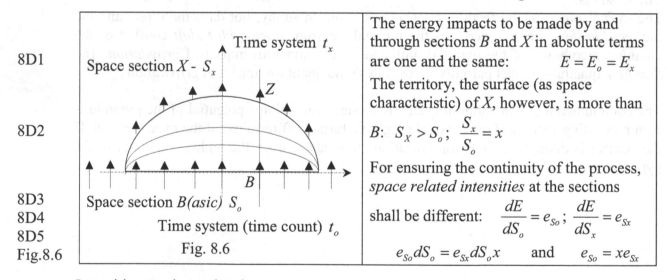

8D1 **8D2** **8D3** **8D4** **8D5** Fig.8.6	Time system t_x Space section X - S_x Z B Space section *B(asic)* S_o Time system (time count) t_o Fig. 8.6

The energy impacts to be made by and through sections *B* and *X* in absolute terms are one and the same: $E = E_o = E_x$

The territory, the surface (as space characteristic) of *X*, however, is more than

B: $S_X > S_o$; $\dfrac{S_x}{S_o} = x$

For ensuring the continuity of the process, *space related intensities* at the sections

shall be different: $\dfrac{dE}{dS_o} = e_{So}$; $\dfrac{dE}{dS_x} = e_{Sx}$

$e_{So}dS_o = e_{Sx}dS_o x$ and $e_{So} = xe_{Sx}$

Intensities are time-related.

The acting intensities of the energy transfer through the surfaces are:

8D6
$$e_o = \frac{de_{So}}{dt_o} = d\frac{dE}{dS_o}\frac{1}{dt_o}; \quad \text{and} \quad e_x = \frac{de_{Sx}}{dt_x} = d\frac{dE}{dS_x}\frac{1}{dt_x} = d\frac{dE}{xdS_o}\frac{1}{dt_x}$$

8D7
8D6 gives the balance for both (and all) sections $e_o\Delta t_o = \dfrac{de_{So}}{dt_o}\Delta t_o = e_{So}$; and $e_x\Delta t_x = \dfrac{de_{Sx}}{dt_x}\Delta t_x = e_{Sx}$

The balance is in order, as the absolute values of the energy transfer through the space sections are equal indeed, value of:

$$E = E_o = e_o \Delta t_o S_o = E_x = e_x \Delta t_x S_x \qquad \text{8D8}$$

With the substitution of 8D5 into 8D7: $\quad e_o \Delta t_o = x e_x \Delta t_x \quad$ and $\quad x = \dfrac{e_o \Delta t_o}{e_x \Delta t_x} = \dfrac{e_{So}}{e_{Sx}} \qquad$ 8D9

From 8D9: $\quad x\dfrac{\Delta t_x}{\Delta t_o} = \dfrac{e_o}{e_x} \quad$ and $\quad x\tau = \dfrac{e_o}{e_x} \quad$ where the time relation of the energy exchange of the surfaces is taken for: $\quad \tau = \dfrac{\Delta t_x}{\Delta t_o} \qquad$ 8E1

The energy *intensity* relation of the two surfaces from 8E1 is: $\quad e_x = \dfrac{e_o}{x\tau} \qquad$ 8E2

8E2 means that, where there is no time count (intensity) difference between the two above the basic and the top surfaces, $\quad \tau = \dfrac{\Delta t_x}{\Delta t_o} = 1 \quad$ the energy intensity relation is established by the relation of the surfaces. $\qquad$ 8D3

Intensity relation means energy transfer in time.

The results in formulas 8D6-8E2 are dimension of $\left[\dfrac{energy}{m^2\ sec}\right]$ but the use of this dimension would only be necessary for calculating the absolute balance of the transfer.

The importance, however, is not about the source, rather the energy intensity impact to and within the quantum systems of reference.

In the case of permanent or continuous energy exchange, the absolute value is not relevant, as it is also not relevant in the case of the proton/neutron balance.

The intensity change characterises the process.

It is necessary to emphasise that while intensity impacts e_{So} and e_{Sx} are from *gravitation*, the relation in 8E2 is not about the difference in the acceleration values of *gravitation*. Ref. 8E2

The acceleration of both spots is taken as $a = g$ equal as both (and any more inside) surfaces are based and taken in motion by the structure – while the impacts are different. For proving this, spots B and Z are taken on the surfaces.

((The value of *gravitation* otherwise is changing by the heights.))

The quantum impacts of *gravitation* at spots B on the basis and Z on the surface are:

The intensity impact at B: $\qquad$ The intensity impact at Z with reference to 8E2 is:

$$e_B = \frac{dmc^2}{dt_i \varepsilon_E}\left(1 - \sqrt{1 - \frac{(g\Delta t)^2}{c^2}}\right) \qquad e_Z = \frac{dmc^2}{x\tau \cdot dt_i \varepsilon_E}\left(1 - \sqrt{1 - \frac{(g\Delta t)^2}{c^2}}\right) \qquad \begin{array}{l}\text{8E4}\\[4pt]\text{8E5}\end{array}$$

$\varepsilon_E = 1$ is the intensity coefficient, $c = 299{,}792$ km/s the quantum speed value of the *Earth*, are taken for both spots, equal.

$[IQ = c_x^2/\varepsilon_x$ - characterises the elementary composition.]

With the acceleration value of *gravitation* $a = g = const$, the intensities of the quantum impacts at e_B and e_Z should be different, otherwise the absolute balance of the impacts would fail.

➤ If $x = 1$ and $\tau = 1$ means: no difference in surfaces, no difference in motion (and in the mineral composition);

the quantum impacts from the surfaces to the quantum systems are equal:

(with reference to the foreword of this section: this is also the proof of the continuity of the quantum impact of *gravitation*) $\qquad e_o = e_x \qquad$ 8F1

➢ If the time related intensities are *equal* to: $\tau = 1$

8F2
as without difference in motion of the basic structure and the surfaces; or
without difference in the mineral composition of the systems of the impact $\quad e_x = \dfrac{e_o}{x}$

➢ If the surface related intensity is *equal* to: $x = 1$

8F3
as with the equality of the surfaces, but with difference in relative motion
or with difference in the elementary (mineral) composition: $\quad e_x = \dfrac{e_o}{\tau}$

➢ $\tau \neq 1$, it means the status of rest/motion conditions of the surfaces are different:

8G1
8G2

either or	either the structure is in relative motion to the basis; or the basis is in relative motion to the structure.
$dt_x = \dfrac{dt_o}{\sqrt{1 - \dfrac{v_x^2}{c^2}}}$; $dt_o = \dfrac{dt_x}{\sqrt{1 - \dfrac{v_x^2}{c^2}}}$ v_x is the speed of the motion.	With reference to 8E2, $\tau > 1$ - the structure is in relative motion, as $\Delta t_x > \Delta t_o$ $\tau < 1$ - the basis is in relative motion, as $\Delta t_x < \Delta t_o$ 8G1 = internal source; 8G2 = external source;

Relative motion indicates: the time systems of the systems of reference above the surfaces are different [also equivalent to the effect of different mineral composition between the surfaces].

The key is c, the speed of quantum communication. It belongs to the system of reference, taken at relative rest.

Gravitation is establishing the quantum system of reference on the surface of the *Earth*. The quantum impact is coming from *gravitation* and the system of rest is belonging to the system of reference of *gravitation* .

➢ $\tau > 1$ means the time system (time-count) within the structure above the basic surface is longer (in motion) and the energy intensity of the system is of less value; the intensity of the energy transmission through its external surface is also of less value;
with $x > 1$ still on, the intensity of the energy transfer is of even less.

➢ $\tau < 1$ and $x > 1$ give opposite results.
(Structures above the *Earth* surface cannot be $x < 1$.)

The general conclusion is that while the absolute values of the energy transfer through the two surfaces are equal: $E_o = E_x$, the acting intensities are different, the accumulating quantum impact is permanent and there is an energy intensity surplus between the two surfaces.

Examples

➢ Structures at relative rest ($\tau = 1$) with $x > 1$ have increased energy intensity content inside.

The best examples are *churches*.

The intensity of *gravitation* and the speed value of quantum communication of a certain spot of the *Earth* surface are of quasi constant values.

[The intensity impact of *Earth* minerals – with specific to the element speed of quantum communication – is to be taken into account and added to the quantum impact, if used in the construction and having been present within the internal space – as the case with the majority of the churches well demonstrates it.]

8H1
Ref.
8D5
The space under the specific architecture of the building is with accumulating increased *gravitation* (and elementary) impact: $\Delta e = e_o - e_x$

With reference to 8D5 the intensity of the incoming *gravitation* impact is more than the intensity leaving through the roof. Ringing bells and organ music increase the impact, as vibration is equivalent to speeding up, which is increasing intensity!
!

➤ Geo-platforms (mountains) at different heights (still with $\tau = 1$)
While the absolute value of the energy impact of *gravitation* is one and the same and the time related intensities are *taken* equal, the intensities of the life process – the utilisation of energy intensities, compared to a flat *Earth* platform without a mountain are different.

With reference to 8F2 and the explanation with 8F4 and 8F5 the quantum impact in the mountains (while the acceleration is strongly constant!) is of less intensity: $e_x = \dfrac{e_o}{x}$ 8H2

as $x > 1$, the surface in the mountains is higher: $S_x > S_o$; resulting in $\Delta e = e_o - e_x$ 8H3

Certain mineral compositions (if resulting in $\tau > 1$) are further increasing the benefit. The less *blue shift* conflict also explains the less measured temperature.   Ref. 8E2

➤ Systems with $\tau \neq 1$ and $x \neq 1$ with reference to 8E2 are with massive intensity difference. Not only the space-, but also the time-related intensities of the components of the systems are different. Systems are either in relative motion to each other or built up from different minerals.
Gravitation is constant but the intensities within the minerals or in a vehicle in motion are different. At low speed values the difference might be difficult to measure.

➤ The intensity impact of mineral columns and specific mineral structures of buildings, acting within the internal quantum system (space) shall be taken into account.
Elementary processes of minerals communicate with increased intensity. This is the result of the increased speed of their quantum communication and the intensity coefficient of their elementary structure. (Reference to the *IQ* value in Section 1.2)   Ref. S.1.2
Earth surface is with $c_{Earth} = 299{,}792$ km/sec quantum speed and electron process intensity coefficient $\varepsilon_E = 1$. Minerals (ores) are with $c_m > 299{,}792$ km/sec and $\varepsilon_m < 1$.
With reference to 8E2 with $x = 1$:

$$\frac{dmc_m^2}{dt_i \varepsilon_m}\left(1 - \sqrt{1 - \frac{(c_m - i_m)^2}{c_m^2}}\right) = \frac{dmc^2}{\tau \cdot dt_i \varepsilon_E}\left(1 - \sqrt{1 - \frac{(c - i)^2}{c^2}}\right);$$ 8H4

or in simple form: $e_x = e_m = \dfrac{e_o}{\tau}$; where $\tau = \dfrac{\Delta t_x}{\Delta t_o} = \dfrac{\Delta t_m}{\Delta t_o} < 1$ equivalent to $\tau \cong \dfrac{IQ_E}{IQ_m} < 1$ 8H5

The relation in 8H5 above is valid, since intensities are established by time relations.
The energy intensity, leaving the mineral column (result of the slower speed of its relative motion) in the case of $x = 1$ would be higher. Columns are however with $x \gg 1$ relation, as their basic surface is less than its cylindrical mantle: they are energy intensity accumulators:

$$e_{in} = \frac{dmc^2}{\tau \cdot dt_i \varepsilon_E}\left(1 - \sqrt{1 - \frac{(c - i)^2}{c^2}}\right); \; e_{out} = \frac{dmc^2}{\tau \cdot x \cdot dt_i \varepsilon_E}\left(1 - \sqrt{1 - \frac{(c - i)^2}{c^2}}\right); \qquad x = \frac{S_{column}}{S_{basis}}$$ 8H6

while $E_o = e_o \Delta t_E = E_m = e_m \Delta t_m = E_{mc} = e_{mc} \Delta t_m x$ 8H7

The transferred impact of *gravitation*, at a surface/level different than the *Earth* surface always represents the actual intensity at that certain level of the measurement.

This is valid for all cases including geo-platforms, where the transferred by the minerals and the hill impact is different than at the flat *Earth* surface; or prove the elementary gaseous composition above the *Earth* surface with $IQ_{O,Ni,He,H} < IQ_E$ expires *gravitation* impact.

The cooling of the elementary processes of the *Earth* by *gravitation*

There is a principal difference between active elementary *components of gravitation* and elementary subjects *taken by gravitation*. Minerals, elementary processes of the *Earth* are the acting drives of the quantum impact of *gravitation*.
Subjects, taken by *gravitation* – while transporting the approaching impact of *gravitation* – are impacting the *Quantum Membrane* by their own anti-electron process *blue shift* impact.

Cooling by *gravitation* means losing on the *blue shift* conflict of the *plasma*, losing on the anti-electron process *blue shift* conflict, giving off in fact the energy intensity of the *plasma*. Elementary processes lose on their anti-electron process *blue shift* surplus, feeding the quantum impact of *gravitation*.
Plasma is electron process *blue shift* conflict of infinite high intensity with infinite high quantum speed.
With the cooling (by *gravitation*) the quantum speed towards the surface is decreasing.
    (Gas status is also *blue shift* conflict but at low speed value of quantum communication.)

$$e_{pl} = \frac{dmc_{pl}^2}{dt_i \varepsilon_{pl}} \left(1 - \sqrt{1 - \frac{(c_{pl} - i_{pl})^2}{c_{pl}^2}} \right) ;$$

with $\lim c_{pl} = \infty$ and $\lim \varepsilon_{pl} = 0$,
with infinite high intensity of quantum
communication towards $\lim c_H = 0$ and $\lim \varepsilon_H = \infty$

8K1

with $\lim \Delta t_n = 0$ and $\lim \varepsilon_n = \infty$

Losing on the infinite high *blue shift* conflict of the *plasma* results in elementary processes with neutron process dominance, a certain balance with the intensity of the proton process.
Plasma is transforming into elementary processes with decreasing quantum speed value, with decreasing electron process intensity and, in this way with decreasing neutron process dominance
Gravitation is
➢ quantum impact of the anti-electron process *blue shift* surplus of the neutron process dominant elementary processes of the *Earth* core, and
➢ "mechanical" *blue shift* impact of the *Quantum Membrane* above the *Earth* surface by the sphere symmetrical expanding acceleration of the *Earth*.

With reference to 1B5, the cooling by *gravitation* is establishing the elementary structures.
The existing *IQ* value of the elementary processes is the drive of their quantum impact potential on the surface of the *Earth*. This is an energy capacity to be utilised.
In the case of the *Uranium* elementary process as mineral for example, this working potential is 37% of the full quantum energy of the elementary process.

Once minerals have been melted and hardened on the *Earth* surface again, their elementary drive becomes reduced, as the speed value of quantum communication is consolidating to the one on the *Earth* surface. The intensity difference of the proton and the neutron processes, characteristic of the element will still be acting, but by significantly less quantum speed value.

"Clean" elements, prepared from minerals have their modified "energy" intensity value.

8K2 *Gold* loses 33% of its quantum communication potential
Silver its 23%,
Chromium its 15%.

$$\left(\frac{c_x^2}{\varepsilon_x} - \frac{c^2}{\varepsilon_x}\right) : \frac{c_x^2}{\varepsilon_x} = \frac{c_x^2 - c^2}{c_x^2}$$

Cooling by gravitation starts from the surface and goes inside.

The schedule here below demonstrates the propagation of the cooling effect within consecutive *Earth* stratums:

e = capacity; $e - \Delta e$ = remaining capacity; Δe = cooling

e	e	e	e		starting *plasma* stage

e		e		e		$e - \Delta e$ Δe	external cooling of
e		e		$e - \Delta e$ Δe		$e - 2\Delta e$ $2\Delta e$	*plasma*
e		$e - \Delta e$ Δe		$e - 2\Delta e$ $2\Delta e$		$e - 3\Delta e$ $3\Delta e$	by *gravitation*
$e - \Delta e$ Δe		$e - 2\Delta e$ $2\Delta e$		$e - 3\Delta e$ $3\Delta e$		$e - 4\Delta e$ $4\Delta e$	starts

...

$e - (n - x)\Delta e$ $(n-x)\Delta e$	...	$e - (n-2)\Delta e$ $(n-2)\Delta e$	$e - (n-1)\Delta e$ $(n-1)\Delta e$	$e - n\Delta e$ $n\Delta e$

Table 8.2

Table 8.2

Table 8.2 above well represents the cooling effect.

Gravitation (cooling) continues until $\lim(e - \Delta e) = 0$, which corresponds to the *Hydrogen* state. *Plasma* is the start, *Hydrogen* process is the end!

In-between is *Life* and the elementary world.

Comment:

The *pyramid* is giving off electron process *blue shift* impact intensity from its conflicting *Quantum Membrane*, at the same time water is taking electron process *blue shift* impact in for increasing its internal conflict by the increase of the quantum speed.

The reason in both cases is the correspondence with the quantum system of the *Earth* surface. Just in the first the need is the release, in the second the taking in.

Printed in the United States
By Bookmasters